VOLUME FIFTY ONE

ADVANCES IN ECOLOGICAL RESEARCH

Big Data in Ecology

ADVANCES IN ECOLOGICAL RESEARCH

Series Editor

GUY WOODWARD
Imperial College London
Silwood Park Campus
Ascot, Berkshire, United Kingdom

VOLUME FIFTY ONE

ADVANCES IN ECOLOGICAL RESEARCH

Big Data in Ecology

Edited by

GUY WOODWARD
Imperial College London
Silwood Park Campus
Ascot, Berkshire, United Kingdom

ALEX J. DUMBRELL
School of Biological Sciences, University of Essex,
Wivenhoe Park, Colchester, Essex, United Kingdom

DONALD J. BAIRD
Environment Canada @ Canadian Rivers Institute,
Department of Biology, University of New Brunswick,
Fredericton, New Brunswick, Canada

MEHRDAD HAJIBABAEI
Biodiversity Institute of Ontario & Department of Integrative
Biology, University of Guelph, Guelph, Canada

AMSTERDAM • BOSTON • HEIDELBERG • LONDON
NEW YORK • OXFORD • PARIS • SAN DIEGO
SAN FRANCISCO • SINGAPORE • SYDNEY • TOKYO
Academic Press is an imprint of Elsevier

Academic Press is an imprint of Elsevier
32 Jamestown Road, London NW1 7BY, UK
525 B Street, Suite 1800, San Diego, CA 92101-4495, USA
225 Wyman Street, Waltham, MA 02451, USA
The Boulevard, Langford Lane, Kidlington, Oxford, OX5 1GB, UK

First edition 2014

Notices
Knowledge and best practice in this field are constantly changing. As new research and experience broaden our understanding, changes in research methods, professional practices, or medical treatment may become necessary.

Practitioners and researchers must always rely on their own experience and knowledge in evaluating and using any information, methods, compounds, or experiments described herein. In using such information or methods they should be mindful of their own safety and the safety of others, including parties for whom they have a professional responsibility.

To the fullest extent of the law, neither the Publisher nor the authors, contributors, or editors, assume any liability for any injury and/or damage to persons or property as a matter of products liability, negligence or otherwise, or from any use or operation of any methods, products, instructions, or ideas contained in the material herein.

ISBN: 978-0-08-099970-8
ISSN: 0065-2504

Printed in the United States of America

CONTENTS

CONTRIBUTORS

Donald J. Baird
Environment Canada @ Canadian Rivers Institute, Department of Biology, University of New Brunswick, Fredericton, New Brunswick, Canada

Mark V. Brown
Evolution and Ecology Research Centre, School of Biological, Earth and Environmental Sciences, University of New South Wales, Sydney, New South Wales, Australia

James M. Bullock
NERC Centre for Ecology and Hydrology, Wallingford, Oxfordshire, United Kingdom

Anthony A. Chariton
CSIRO Oceans and Atmosphere, Lucas Heights, New South Wales, Australia

Steve Cinderby
Environment, University of York, Heslington, York, United Kingdom

Katherine A. Dafforn
Evolution and Ecology Research Centre, School of Biological, Earth and Environmental Sciences, University of New South Wales, Sydney, and Sydney Institute of Marine Sciences, Mosman, New South Wales, Australia

Isabelle Durance
Cardiff School of Biosciences, Cardiff, United Kingdom

Bridget Emmett
NERC Centre for Ecology & Hydrology, Environment Centre Wales, Bangor, Gwynedd, United Kingdom

Stephanie Gardham
Department of Environment and Geography, Macquarie University, Sydney, and CSIRO Oceans and Atmosphere, Lucas Heights, New South Wales, Australia

Jim Harris
Environmental Science and Technology Department, School of Applied Sciences, University of Cranfield, Cranfield, United Kingdom

Kevin Hicks
Environment, University of York, Heslington, York, United Kingdom

Grant C. Hose
Department of Biological Sciences, Macquarie University, Sydney, New South Wales, Australia

Emma L. Johnston
Evolution and Ecology Research Centre, School of Biological, Earth and Environmental Sciences, University of New South Wales, Sydney, and Sydney Institute of Marine Sciences, Mosman, New South Wales, Australia

Brendan P. Kelaher
National Marine Science Centre, Southern Cross University, Coffs Harbour, New South Wales, Australia

Tom H. Oliver
NERC Centre for Ecology and Hydrology, Wallingford, Oxfordshire, United Kingdom

Dave Paterson
Scottish Oceans Institute, East Sands, University of St. Andrews, St. Andrews, Scotland, United Kingdom

Dave Raffaelli
Environment, University of York, Heslington, York, United Kingdom

Stuart L. Simpson
CSIRO Land and Water, Lucas Heights, New South Wales, Australia

Sarah Stephenson
CSIRO Oceans and Atmosphere, Lucas Heights, New South Wales, Australia

Melanie Y. Sun
Evolution and Ecology Research Centre, School of Biological, Earth and Environmental Sciences, University of New South Wales, Sydney, and Sydney Institute of Marine Sciences, Mosman, New South Wales, Australia

Piran C.L. White
Environment, University of York, Heslington, York, United Kingdom

PREFACE

Guy Woodward*, Alex J. Dumbrell[†], Donald J. Baird[‡], Mehrdad Hajibabaei[§]
*Imperial College London, United Kingdom
[†]University of Essex, United Kingdom
[‡]Environment Canada, Canada
[§]University of Guelph, Canada

Ecology is entering previously uncharted waters, in the wake of the huge growth in "Big Data" approaches that are beginning to dominate the field. Previously, the rate at which ecology advanced, especially when dealing with large scales and multispecies systems, was limited by the paucity of empirical data, which was often collected in a painstaking and labour-intensive manner by a few dedicated individuals. We are now entering a phase where the polar opposite situation is the norm and the new rate-limiting step is the ability to process the vast quantities of data that are being generated on an almost industrial scale and, more importantly, to interpret their ecological significance. This ecoinformatics revolution is happening simultaneously on many fronts: from the exponential increases in sequencing power using novel molecular techniques, to the increased capacity for remote sensing and high-resolution GIS, and the marshalling of huge volumes of metadata collected by both the scientific community and the rapidly swelling ranks of Citizen Scientists. This latter group will account for a sizeable portion of the Big Data that needs to be handled in future: Citizen Scientists are already starting to eclipse the capacity of official bodies to carry out large-scale and long-term routine data collection and biomonitoring, as the traditional boundaries between natural and social sciences and data ownership become evermore blurred. This democratisation and sharing of data among scientists, across disciplines, and with the lay public that has gone hand in hand with Big Data approaches is altering the very nature of scientific discourse in a profound manner, and in ways that we do not yet fully comprehend. This volume highlights three examples of some of the main Big Data trends and their potential to address the big questions in ecology in this new multidisciplinary era.

In addition to geospatial data series and large federated databases that are becoming commonplace, particularly in the field of biomonitoring and remote sensing, ecogenomics represents both one of the greatest informatics resources and one of the biggest emerging challenges in ecology. This is a rapidly growing field, and the recent explosion of molecular ecology embraces a plethora of terms that were barely on the horizon a decade ago, including metasystematics, metranscriptomics, and functional genomics, among others. These terms are entering the day-to-day lexicon of ecologists at an accelerating rate, and they are now frequently seen in both grant proposals and peer-reviewed publications. Even so, most ecological studies that use such approaches are still restricted to descriptive "fishing expeditions", rather than being used for explicit hypothesis generation or testing. Thus, although countless recent papers have revealed previously unguessed-at levels of biodiversity in even the most remote and hostile environments, particularly in the microbial world, very few have been couched in the rigorous hypothetico-deductive framework that is the bread and butter of the more established fields of mainstream ecology. In the light of this, it is critically important that in the heady rush to adopt Big Data approaches, we must take care to corroborate them with more traditional techniques, if only to enable a degree of handshaking before jettisoning obsolete technologies: otherwise, we run the risk of creating a schism in ecology that could lead to huge inefficiencies in the future, where we simply end up asking the same old questions but with different data, rather than truly advancing the field.

Before ecogenomics techniques and data are widely applied, they must therefore first provide credible evidence that they can do at least what existing techniques can do, but with added value. In the paper by Dafforn et al. (2014), the authors describe a case study that applies a metagenomics approach in estuarine ecosystems in Australia, while comparing the results with a parallel approach using traditional taxonomic analysis. The authors demonstrate convincingly that, despite the bioinformatics challenges, the ecogenomics approaches clearly provide data far more rapidly and efficiently, with benthic assemblages resolved at higher levels of taxonomic resolution. Perhaps even more importantly, though, they provide far stronger insights into the major environmental drivers of composition across a range of contrasting estuarine ecosystem conditions. In the second paper in the volume, by Gardham et al. (2014), a comparable metagenomics approach is applied to analyse mesocosm experiments studying the effects of metal pollution on freshwater benthic assemblages. When focused on the microbial community in particular, the exploratory power of multivariate approaches is greatly enhanced, in terms of exploring assemblage pattern-driver relationships, and this offers a huge new potential means of ecological indicator development. While metagenomics approaches are now being more widely applied in ecosystem research, both studies illustrate the opportunities created through the application of these new techniques, and also the emergence of the new generation of studies that are starting to embed Big Data into more explicitly hypothesis-focused frameworks. They also illustrate how Big Data processing requirements make it more crucial than ever to understand the complex analytical pathways that turn terabytes of DNA sequence into trustworthy ecological information.

The mushrooming of such sequence-based databases provides a vast and potentially invaluable resource for current and future generations of ecologists (Fig. 1), but increasingly concerns have been raised about the stringency of quality assurance and ground-truthing of the underlying data, which could seriously undermine the field if errors are being propagated unwittingly and repeatedly and on a potentially grand scale: i.e. there can be a world of difference between Big Data and Good Data. Notwithstanding

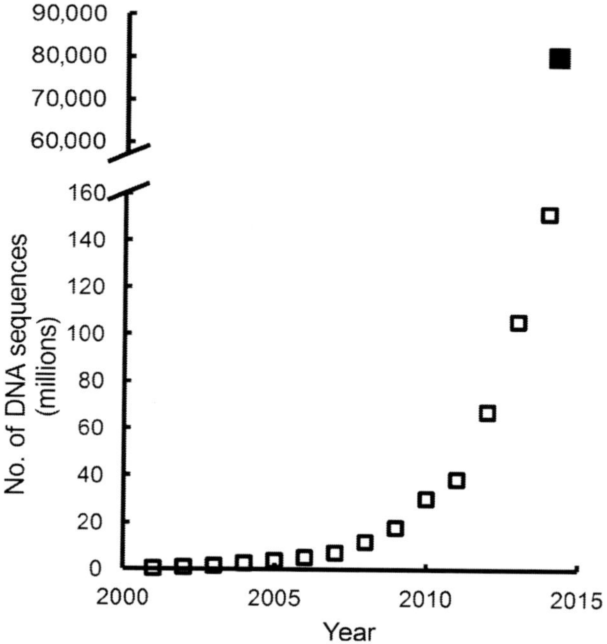

Figure 1 The number of DNA sequences contained within the GenBank database (the principal non-NGS sequence repository) as a function of time (open symbols). This acts as a proxy for publication quantity as you can't publish DNA sequences without first providing them to GenBank. These data include non-ecological DNA sequences. The solid symbol at the top is current number of DNA sequences contained within the MG-RAST repository, which only stores metagenome data, i.e. whole-community ecological data. Note the change in axis scales and how metagenomic approaches over the course of a couple of years has now produced more DNA sequences than the entire GenBank collection.

these underlying issues, the rate of data generation that can now be achieved at relatively little cost is breathtaking and would have been inconceivable just a few years ago. It is also the sophistication of the data and the fact that multiple forms of information are being synthesised and compiled simultaneously that form the hallmarks of the most recent advances in this area. Collated databases containing outputs from multiple ecological studies will soon surpass single studies in terms of data breadth, and emerging molecular (e.g. next-generation sequencing) approaches will dwarf other ecological data in terms of depth and breadth of coverage of multispecies systems: in fact, it could be argued that this revolution has already happened (Fig. 1).

There is another major source of large ecological datasets that are becoming increasingly prevalent, which also present associated Big Data challenges, and this comes in the form of the outputs of large-scale multi-institutional (often multi-national) research programmes. Within the UK, the Natural Environment Research Council recently launched the Biodiversity and Ecosystem Service Sustainability Programme (BESS; 2011–2017), a multimillion pound investment that represents a UK-wide effort to characterise the links between biodiversity stocks and flows of ecosystem services across a broad spectrum of terrestrial and aquatic landscapes (http://www.nerc-bess.net/). This ambitious programme is led by Professor Dave Raffaelli (University of York), and the paper he leads in this volume (Raffaelli et al., 2014) highlights the Big Data challenges faced by BESS and the approaches being used to overcome these. Raffaelli et al. begin with lessons that can be learnt from history and draw the readers' attention to the pioneering International Biological Programme (IBP), which ran from 1964 to 1974 and was one of the first to attempt what we now call Big Data ecology. The IBP was in many ways too far ahead of its time, and it was beset by numerous problems resulting from its own huge complexity and scale of ambition, and it was abandoned long before its full potential could be realised. Raffaelli et al. highlight how half a century later we are only now finally starting to be able to deal with the size and scope of this style of research programme. It is only in the last few years that we have been able to wield the necessary tools for such a complex and challenging undertaking, and these were unfortunately lacking in the 1960s. To illustrate this, Raffaelli et al. explore the different approaches taken by the four main projects within BESS, which work to answer similar ecological questions, but in very different systems: remote upland streams, lowland agricultural landscapes, urban areas, and coastal environments. They then demonstrate how data from each of these can be integrated before looking to the future to address emerging challenges as the datasets continue to expand in both volume and scope. This form of large-scale and multidisciplinary research programme is increasingly becoming the norm, and indeed, it is a prerequisite for many research funding schemes, especially in Europe, as it is widely seen as being essential for understanding and predicting the behaviour of seemingly complex ecosystems in the human-dominated twenty-first century. The days of the lone researcher working in splendid isolation on a narrowly focussed problem are fading fast, as the need to develop broad collaborations that span traditional disciplinary boundaries means that "science by committee" has become the norm in the age of Big Data: this is especially true at the interface of the natural and social sciences, where the impacts of humans on ecosystem services have become a huge focus of research activity in a matter of just a few years. Whether this fundamental shift in the way ecology is conducted is entirely healthy is a question that merits further debate, as there is a real danger that the gifted auteurs that have previously driven many of the field's biggest advances may be left behind in this very different future landscape. Nonetheless, it seems inevitable that at least in the foreseeable future, the impetus will continue to be with ambitious, large-scale science, as the renaissance of the IBP's legacy continues to gather strength, underpinned by advances in Big Data. Given the rapidly accelerating rate at which ecology is now progressing, it seems certain that dramatic revolutionary advances lie ahead in the near future that we cannot yet even imagine, and we hope that this volume helps to move us a little further and a little faster forwards towards that goal.

REFERENCES

Dafforn, K.A., Baird, D.J., Chariton, A.A., Sun, M.Y., Brown, M., Simpson, S.L., Kelaher, B.P., Johnston, E.L., 2014. Faster, higher and stronger? The pros and cons of molecular faunal data for assessing ecosystem condition. Adv. Ecol. Res. 51, 1–40.

Gardham, S., Hose, G., Stephenson, S., Chariton, A., 2014. DNA metabarcoding meets experimental ecotoxicology: advancing knowledge on the ecological effects of copper in freshwater ecosystems. Adv. Ecol. Res. 51, 79–104.

Raffaelli, D., Bullock, J.M., Cinderby, S., Durance, I., Emmett, B., Harris, J., Hicks, K., Oliver, T.H., Paterson, D., White, P.C.L., 2014. Big data and ecosystem research programmes. Adv. Ecol. Res. 51, 41–78.

CHAPTER ONE

Faster, Higher and Stronger? The Pros and Cons of Molecular Faunal Data for Assessing Ecosystem Condition

Katherine A. Dafforn[*,†,1], Donald J. Baird[‡], Anthony A. Chariton[§], Melanie Y. Sun[*,†], Mark V. Brown[*], Stuart L. Simpson[||], Brendan P. Kelaher[¶], Emma L. Johnston[*,†]

[*]Evolution and Ecology Research Centre, School of Biological, Earth and Environmental Sciences, University of New South Wales, Sydney, New South Wales, Australia
[†]Sydney Institute of Marine Sciences, Mosman, New South Wales, Australia
[‡]Environment Canada @ Canadian Rivers Institute, Department of Biology, University of New Brunswick, Fredericton, New Brunswick, Canada
[§]CSIRO Oceans and Atmosphere, Lucas Heights, New South Wales, Australia
[¶]National Marine Science Centre, Southern Cross University, Coffs Harbour, New South Wales, Australia
[||]CSIRO Land and Water, Lucas Heights, New South Wales, Australia
[1]Corresponding author: e-mail address: k.dafforn@unsw.edu.au

Contents

Advances in Ecological Research, Volume 51
ISSN 0065-2504
http://dx.doi.org/10.1016/B978-0-08-099970-8.00003-8

Abstract

Ecological observation of global change processes is dependent on matching the scale and quality of biological data with associated geophysical and geochemical driver information. Until recently, the scale and quality of biological observation on natural assemblages has often failed to match data generated through physical or chemical platforms due to constraints of cost and taxonomic resolution. With the advent of next-generation DNA sequencing platforms, creating 'big data' scale observations of biological assemblages across a wide range of phylogenetic groups are now a reality. Here we draw from a variety of studies to illustrate the potential benefits and drawbacks of this new data source for enhancing our observation of ecological change compared with traditional methods. We focus on a key habitat—estuaries—which are among the most threatened by anthropogenic change processes. When community composition data derived using morphological and molecular approaches were compared, the increased level of taxonomic resolution from the molecular approach allowed for greater discrimination between estuaries. Apart from higher taxonomic resolution, there was also an order of magnitude more taxonomic units recorded in the molecular approach relative to the morphological. While the morphological data set was constrained to traditional macroinvertebrate sampling, the molecular tools could be used to sample a wide range of taxa from the microphytobenthos, e.g., diatoms and dinoflagellates. Furthermore, the information provided by molecular techniques appeared to be more sensitive to a range of well-established drivers of benthic ecology. Our results indicated that molecular approaches are now sufficiently advanced to provide not just equivalent information to that collected using traditional morphological approaches, but rather an order of magnitude bigger, better, and faster data with which to address pressing ecological questions.

1. INTRODUCTION

1.1. Bioassessment and monitoring of ecosystem change

The ecological measurement of global change processes is dependent on matching the scale and quantity of biological data with associated geophysical and geochemical driver information (Baird and Hajibabaei, 2012). Until recently, the scale and quality of biological observation on natural assemblages has failed to match data generated through physical and chemical platforms due to constraints of cost and taxonomic resolution (Friberg et al., 2011). With the advent of next-generation DNA sequencing platforms, generating 'big data' scale observations on biological assemblages across a wide range of phylogenetic groups is now a reality (Baird and Hajibabaei, 2012; Brown et al., 2009; Chariton et al., 2010a; Hajibabaei et al., 2011; Kohli et al., 2014; Sogin et al., 2006; Sun et al., 2013). Big data can be defined as large volumes of data that require novel data processing tools and strategies (Hampton et al., 2013). Dealing with big data can be challenging, but presents great opportunity for data-intensive biomonitoring approaches. Here, we illustrate the potential advantages of this new data source in observation of ecological change, illustrating the pros and cons of this new approach, focusing on estuaries which are among the most anthropogenically disturbed marine habitats (Kennish, 2002).

Observing natural ecosystems, particularly at large scales, requires a consistent approach to data collection (Birk et al., 2013). A major current constraint is the necessity of limiting the phylogenetic breadth of observation to what is practical in terms of timely data generation (Friberg et al., 2011). For this reason, studies have tended to converge on particular groups of well-studied and taxonomically tractable species (e.g. fish, macroinvertebrates) (Chariton et al., 2010b; Dafforn et al., 2012, 2013; McKinley et al., 2011), which are characterised by ease of collection and identification, as well as their importance to industry (e.g. fisheries) and ecosystem processes. However, despite the widespread collection of such data in ecological studies and environmental monitoring programmes, data integration to link common responses across taxonomic groups remains challenging.

Biomonitoring science focuses on using patterns in the occurrence and characteristics of individual taxa and/or biological assemblages to interpret ecological change. This normally takes the form of simple binary analysis (divergent/non-divergent), or a 'shades of grey' classification. Most biomonitoring programmes employ sets of phylogenetically constrained observations to bolster a lack of comprehensive biological coverage. For example, in river monitoring, separate sampling approaches are employed to study fish, macroinvertebrates and attached algae (periphyton) (Birk et al., 2012; Bonada et al., 2006). Generally, these observational approaches have evolved in parallel, but inevitably suffer from divergent spatio-temporal sampling approaches and the amount of cost and effort expended to obtain samples (Cao and Hawkins, 2011). Their compatibility for integrated analysis of ecosystem-level change is therefore questionable. It should also be noted that this incompatibility is also driven by the vagaries arising from parallel research traditions resulting in divergent communities of scientific practice.

Observing ecological change requires careful and clear formulation of research questions. For example, Magurran et al. (2010) noted that the spatial and temporal properties of the observation units (e.g. taxa groups, habitat units) should be pertinent to the question being asked. For example, if a migratory species is being studied, it is important to ensure that the species is present when seasonally intermittent stressors are the subject of study in a specified habitat area. Moreover, with an increased focus on improved observational quality in terms of taxon occurrence at local scales, with increased frequency (e.g. as suggested by Harris and Heathwaite, 2012), then separation of driver–response signals from noise should be possible, and, ideally, quantifiable in either an absolute or probabilistic sense (Baird and Hajibabaei, 2012).

Comparisons of relevant chemical contaminant concentrations and ecological health measures across estuaries are challenging, due to large natural variation. To overcome this, comparisons over multiple estuaries require substantial spatial and temporal replication to provide adequate statistical power to detect human impacts (Underwood, 1991). Recent efforts to monitor these impacts have focused on integrating information collected from chemical and ecological monitoring into a more holistic understanding of estuarine condition (Borja et al., 2008; Chariton et al., 2010b; Dafforn et al., 2012). However, we still lack quantitative information at multiple scales, which can be summarised for comparison across whole estuaries or coastal regions, which are essential if they are to be broadly implemented for assessment and management purposes.

In situations where prevailing environmental drivers/stressors are manifold, and where there is a desire to separate specific drivers, it is useful to have rich taxonomic information to allow discrimination (Baird and Hajibabaei, 2012; Burton and Johnston, 2010; Olsgard et al., 1998). However, biological observations remain constrained by a general focus on limited phylogenetic groupings due to the difficulties of obtaining high-resolution taxonomic information. Thus, the interpretation of patterns observed at ecosystem scales are necessarily constrained to a limited number of 'observable receptors', leading to weak inference. Moreover, when cross-ecosystem comparisons are being made, it is valuable to clearly separate system-specific patterns manifested at the local community scale from those occurring at the metacommunity scale (Heino, 2013). For this reason, increasing the number of 'observable receptors' is one potential route towards the development of stressor-specific diagnostic responses at ecosystem scale: the ability to observe hundreds to thousands of entities offers greater potential to observe unique, taxon-stressor responses which can be aggregated and interpreted at the assemblage scale (see Fig. 1.1, for further details). Moreover, analysing the relative contribution of multiple drivers to biological patterns observed at the ecosystem scale using multivariate statistics (Friberg et al., 2011; Lücke and Johnson, 2009) is constrained by the number of simultaneous observations, which are available to include in the analysis. Where these are similar in magnitude to the number of driver

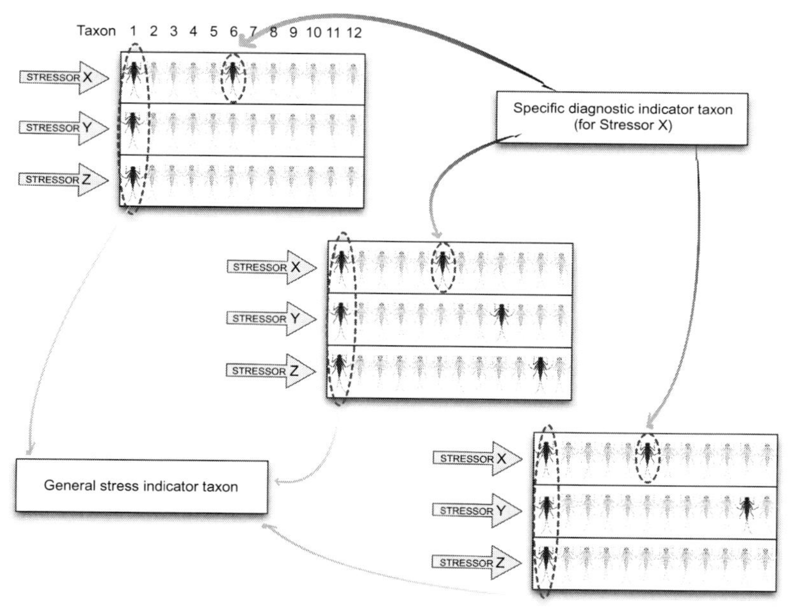

Figure 1.1 Boxes represent the responses of hypothetical taxon assemblages (e.g. from survey samples and mesocosm experiments) to multiple stressors X, Y and Z. In each case, taxa responding are indicated in red (dark grey in print version); those not responding are indicated in green (light grey in print version). One taxon responds equally to all stressors—and can be classed as a 'general stress indicator', but its indiscriminate response provides no diagnostic value. On the other hand, one taxon clearly responds only to Stressor X and can be classed as a 'potential diagnostic indicator of Stressor X'. Following this logic, expanding the range of taxa (=receptors) increases the likelihood that diagnostic indicator taxa can be identified and thus add diagnostic value to ecological assessments.

variables, over-fitting can result in potential erroneous inferences regarding driver–assemblage responses (Green, 1991; Quinn and Keough, 2002). A method is therefore needed which can generate large numbers of consistent observations of taxon occurrence, thus moving the diagnosis of cause in multiple stressor scenarios towards big data approaches (Woodward et al., 2014).

1.2. Application of molecular tools in biomonitoring

With the advent of high-throughput sequencing platforms, it is now possible to consider a comprehensive analysis of the biological structure of environmental samples (Creer et al., 2010; Shokralla et al., 2012; Zinger et al., 2012). By using a combination of multiple gene markers, carefully selected primers and a dedicated bioinformatics pipeline, it is possible to generate a more phylogenetically complete snapshot of the biodiversity of a community (Coissac et al., 2012; Hajibabaei et al., 2011; Morgan et al., 2013; Stoeck et al., 2010).

Baird and Hajibabaei (2012) introduced the concept of 'Biomonitoring 2.0' to describe the shift towards causal analysis in ecological assessment. A key tenet of this approach is that the increase in the numbers of taxa observed using DNA-based molecular identification results in a similar increase in the numbers of 'receptors' responding to specific sets of environmental variables. In this way, the step-change in the numbers of unique 'receptor-entities' offers significant potential for statistical discrimination of cause (Fig. 1.1).

1.3. Assessing estuarine condition

Estuaries can be broadly defined as the interface between fresh and marine waters (Kennish, 2002). These systems are inherently spatially and temporally complex, making the development of protocols founded on predictability, e.g., routine biomonitoring programmes, challenging (Akin et al., 2003; Chapman and Wang, 2001). At the semi-diurnal scale, large variations in the physico-chemical properties of the overlying waters can occur with the ebb and flow of the tides, with the extent of these variations being driven by many factors, including channel and mouth morphology, tidal regime and distance from the mouth. In high energy areas, the overlying waters may shift from being marine to freshwater dominated (Rogers, 1940). In addition, expanses of inter-tidal sediments may be directly exposed to the air during the lower phase of tide. Such large and rapid changes in environmental conditions undoubtedly place considerable physiological pressure on estuarine residing biota (Elliott and Quintino, 2007). This is especially the case when considering the impact of a rapid change in salinity, where organisms have developed a range of behavioural and physiological strategies to cope with such challenges (Charmantier et al., 2001). Clearly, such adaptations are not universal, and the biological diversity within the more physiologically challenging areas (e.g. poikilohaline areas where salinity variation is of biological significance) is generally lower than that of more stable areas, such as the predominately marine waters (euhaline) at the front of estuaries. The underpinning view is that for both macro (>500 μm) and meiofauna (0.1–500 μm) biological diversity is appreciably lower in low salinity environments (Reizopoulou et al., 2013). However, in contrast to macrofauna, meiofaunal biomass does not decline with salinity (Remane, 1934). Changes in biological compositions of estuaries are not solely driven by salinity, with large variations in community composition observed along gradients of sediment grain size, nutrients and organic material loading (Chariton et al., 2010b; Dafforn et al., 2013; Elliott and Quintino, 2007).

The ecology of estuaries is also driven by marked changes in environmental conditions which occur over more protracted periods. The most obvious of these is seasonal variation, which can force massive changes in productivity and biomass particularly in temperate systems (e.g. Kelaher and Levinton, 2003). Other examples include periods of high rainfall and freshwater inflow that can limit the influence of euhaline and even brackish waters to the mouth of the estuaries. Conversely, during drier periods, the influence of saltwater may extend further upstream. In common with rivers and lakes, the physico-chemical and biological characteristics of an estuary are strongly shaped by the surrounding catchment and its land-use. With approximately 60% of the human population residing within 100 km of the coast (Vitousek et al., 1997), the ecological foot print of anthropogenic activities on estuarine system is often marked. The primary direct and indirect anthropogenic stressors vary greatly across systems, and often include a range of point and diffuse sources. For many systems, the key stressors include alterations in the proportions of fresh and marine waters due to water extraction and changes in mouth morphology; eutrophication from excess nutrients; over harvesting of commercial species; increased rates of sedimentation due to run-off and a loss of riparian vegetation, seagrass beds and mangrove stands; as well as contaminants, including legacy contaminants which persist due to their absorption to the sediments (Kennish, 2002). In addition, environmental stressors associated with climate change, e.g., decrease in pH and an increase in saltwater intrusion, are also becoming increasing apparent (Elliott et al., 2014; Kennish, 2002). For scientist and environmental managers, one of the great challenges is being able to identify whether changes in biological communities and ecosystem processes are being driven by natural phenomena, specific anthropogenic activities, or a combination of both (Elliott and Quintino, 2007).

While it is apparent that anthropogenic contaminants such as metals (e.g. Cd, Cu, Pb, Zn) and organics (e.g. polycyclic aromatic hydrocarbons (PAHs)) have an impact on benthic communities (Burton and Johnston, 2010), there remain great challenges in quantifying the degree of impact caused by individual contaminants or even class of contaminants (metals, organics). A frequent outcome of benthic ecology studies with matching environmental contaminants data is that, in combination but not individually, increased contaminant concentrations often explain a large portion of the ecological change. This occurs because the concentrations of many of the contaminants and physico-chemical factors that increase the accumulation of contaminants (e.g. particle size and organic carbon) are strongly correlated.

The more comprehensive ecological data sets provided by molecular tools may potentially allow for greater discrimination of the effects of individual contaminants.

1.4. Case study: Contrasting molecular big data with traditional morphological tools

Next-generation DNA sequencing platforms allow us to generate "big data" scale observations of biological assemblages, but the advantages of these techniques over traditional morphological tools require detailed analysis. Rarely are observational studies designed to comprehensively co-sample for both sequencing and morphological analyses (e.g. Chariton et al., 2014; Gardham et al., 2014). We used co-sampled sediments from a large-scale field study of estuary health to assess the advantages of new molecular techniques over traditional morphological tools for ecological observation. Different techniques to quantify ecological impact have utilised changes to community composition as well as changes to diversity and abundance with the prediction that negative effects of anthropogenic contaminants would manifest themselves as compositional changes or reductions in species diversity, potentially indicating reduced function (Chariton et al., 2010a). Indeed Johnston and Roberts (2009) found that species richness was reduced by and average of ~40% across a range of contaminated marine systems compared to reference sites. Here, we compare traditional morphological data against molecular sequencing data with a variety of indices commonly used to examine estuarine condition. These were

(1) the sediment community composition (a) sub-sampled to include only taxa found using both approaches and pooled to the same level of taxonomic resolution; (b) including all taxa identified and analysed at the highest taxonomic level;

(2) the relationships among the sediment community identified using each approach and a variety of individual and grouped anthropogenic stressors; and

(3) the richness of individual taxa, polychaete families and crustacean orders.

2. METHODS

2.1. Estuarine survey design

Field surveys in multiple estuaries were used to compare information provided by molecular techniques with that provided by traditional morphological techniques for assessing benthic sediment health. Six sites (between 1 and 2 km apart) were sampled from each of eight estuaries along the coast of New South Wales, Australia (Fig. 1.2). Port Kembla, Hunter River, Port Jackson and Georges River are urbanised estuaries with histories of industrialisation. Hacking River, Clyde River, Hawkesbury River and Karuah River are estuaries that are relatively less modified by urbanisation and have no history of major industry. Furthermore, Clyde River estuary is a Marine Protected Area, and sites in Hacking River and Karuah River were also in, or adjacent to, Marine Protected Areas (Fig. 1.2).

2.2. Benthic sediment sampling

Benthic sediments were collected subtidally (~5 m depth) between February and March 2011 using a Van Veen sediment grab. Three sediment grabs were collected at each site and sub-sampled for the surface microbial community (<1 cm depth) and for chlorophyll-a analysis using separate sterile 50-mL Falcon tubes. Each grab sample was homogenised in a clean tray and sub-sampled for infauna community analysis using a 250-mL plastic jar. Sub-samples were also collected to assess anthropogenic contamination (metals and PAHs) and organic enrichment (total organic carbon (TOC) and silt content (% <63 μm)). Plasticware used to collect sediment for metals

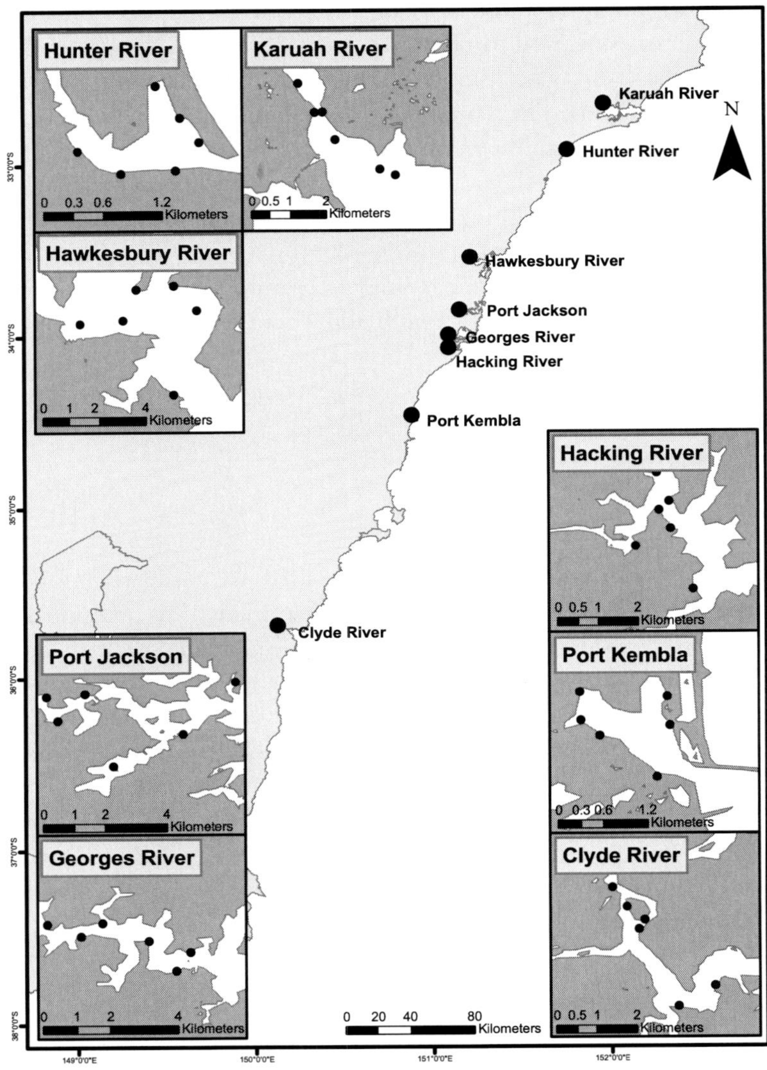

Figure 1.2 Map of study sites along the New South Wales coastline, SE Australia. Port Kembla, Hunter River, Port Jackson and Georges River are heavily modified estuaries. Karuah River, Hawkesbury River, Hacking River and Clyde River are relatively unmodified estuaries.

analyses was previously soaked in 5% HNO_3 for a minimum of 24 h and then rinsed in deionised water (Milli-Q™). Samples were kept in the dark on ice for transport to the laboratory and then samples for chemical analyses were frozen at −20 °C. Details of chemical analyses are included in

Appendices A–C. Sediment deposition was estimated from a sediment trap (30 × 5 cm Perspex cylinders) deployed at each site for 3 months.

2.3. Morphological biomonitoring

Infaunal sub-samples (125-mL) were stained with Rose Bengal and preserved in a 7% formalin solution then passed through a 2-mm mesh (to remove large debris) and onto a 500-μm sieve. The remaining organisms were sorted with a dissecting microscope and identified to the lowest feasible taxonomic level (mostly order for the crustaceans or family for the polychaetes). A reference collection was deposited at the Australian Museum.

2.4. Molecular biomonitoring

Total genomic DNA was extracted from 8 g of each surface sediment sample ($n = 144$) using the PowerMax™ Soil DNA Isolation Kit (Mo Bio Laboratories Inc., Carlsbad, CA, USA). Eukaryotic microbial community composition was determined using 454 ribosomal tag pyrosequencing targeting 18S rRNA genes. 18S primers all18SF (5′-TGGTGCATGGCCGTTCTTAGT-3′) and all18SR (5′-CATCTAAGGGCATCACAGACC-3′) were used to amplify between 200 and 500 base pair product corresponding to the 18S rRNA gene-v9 hypervariable region (Hardy et al., 2010). Amplicons were sequenced on a whole plate by the Australian Genome Research Facility Ltd. (Brisbane, Australia) using a Roche GSFLX pyrosequencer with short-read chemistry (Roche Applied Science, Indianapolis, IN, USA).

Sequences were filtered for mismatching primers, ambiguous bases, homopolymers and low-quality score windows using QIIME to ensure sequence fidelity (Caporaso et al., 2010) and 1 080 222 quality 18S reads remained. Using the Usearch algorithm for *de novo* chimaera detection <1% of 18S sequences were identified as chimaeras and discarded. Operational taxonomic units (OTUs) were aligned and assigned species level taxonomy using the SILVA small subunit ribosomal v115 release for 18S sequences (method: BLAST). Clustering was performed at a sequence similarity cut-off of 97% for OTU generation (Huse et al., 2010). The number of OTUs was sub-sampled to standardise sequencing effort (1828 18S OTUs) leaving a total of 10,857 unique 18S OTUs. Sequences unclassified at the Phylum level were removed. Finally, rare sequences with <4 occurrences or those found in only one sample were discarded leaving a final 18S data set of 3575 OTUs.

To determine potential cross contamination between samples and the appropriateness of the chosen OTU generation, two control assemblages of seven to eight clone sequences were treated alongside samples during the amplicon library and barcode amplification steps. Both control assemblages were then sequenced alongside the samples, with one control in each gasket. Results revealed minimal cross contamination between samples, and data indicates that clustering at 97% similarity was realistic but slightly overestimated diversity, with the 15 clones identified as 22 unique OTUs, a result comparable to other studies of microbial diversity and composition.

2.5. Measuring anthropogenic stressors

Benthic sediments collected for metal analyses were oven dried at 50 °C for 24–48 h and oyster tissue was freeze-dried for 48 h before being homogenised to a fine powder in a ball mill (Retsch, GmbH-301 mm, Germany). Analyses of total recoverable metal concentrations of benthic and suspended sediments were made using high temperature microwave-assisted (MARS 5, CEM) digestion methods, whereby 0.5 g of dry sediment was digested in 9 mL HNO_3 and 3 mL HCl for 18 h cold, then 4.5 min at 175 °C. Metal concentrations in the final digest solutions were analysed using inductively coupled plasma-atomic emission spectrometry (ICP-AES, Varian730 ES). For quality assurance, analyses were made of acid-digest blanks, replicates for >20% of samples, analyte sample-spikes, and PACS-2 certified reference material (CRM). Replicates were within 20% and recoveries for spikes and sediment CRM for metals were within 85–99% of expected values. The limits of reporting for the various methods were less than 1/10th of the lowest measured values.

PAHs analysed were naphthalene (Nap), acenaphthylene (Acel), acenaphthene (Ace), fluorene (Flu), phenanthrene (Phe), anthracene (Anth), fluoranthene (FluA), pyrene (Pyr), benz(a)anthracene (BaA), chrysene (Chry), benzo(a)pyrene (BaP), benzo(b)fluoranthene (BbF), benzo(k) fluoranthene (BkF), indeno(1,2,3-cd)pyrene (Ind), dibenzo(a,h)anthracene (DahA) and benzo(g,h,i)perylene (BghiP). Analyses of PAHs in sediments followed Method 8260 (USEPA, 1996). Surrogate PAHs (deuterated internal standards; acenaphthene-d_{10}, phenanthrene-d_{10}, chrysene-d_{12} and perylene-d_{12}) were spiked into all samples and recoveries were $111 \pm 19\%$. PAH concentrations were normalised to 1% TOC for comparison with sediment quality guideline values (ANZECC/ARMCANZ, 2000).

Inorganic carbon in benthic sediments was removed by acidification with 2 mL of 1 M HCl overnight (Hedges and Stern, 1984), and TOC was analysed using a Leco CN2000 analyser (Leco Corporation, USA) at a combustion temperature of 1050 °C.

Sediments for chlorophyll-a analysis were freeze-dried for 48 h before being homogenised with a stirrer. To ~3 g (accurately weighed) sediment, 30 mL of 90 % acetone was added and samples vortexed before being placed into a sonicator for 15 min. Samples were refrigerated for 18 h to steep before being vortexed and sonicated for a second time and finally centrifuged at 8000 rpm for 10 min to remove turbidity. Samples were analysed in a spectrophotometer (Lambda 35 UV/vis Spectrophotometer) using an acidification technique to determine chlorophyll-a concentrations. Two blank samples consisting of 3 mL of 90% acetone were placed in a 1 cm path-length cuvette and run at the beginning and end of each session, in addition to one blank after every 12 samples. For each sample the equipment was pre-rinsed, and 3 mL of extract was placed within the cuvette. The absorbance was recorded at 630, 647, 664, 665, 691 and 750 nm before acidification with 0.1 mL of 0.1 M HCl. Following the addition of the acid the cuvette was gently agitated and the absorbance remeasured after 90 s (Greenberg et al., 1992).

Sediment grain size analyses (one to two replicates per site randomly selected) were made by wet sieving through stainless steel sieves; gravel (2 mm), sand (2 mm to 63 μm) and fines (<63 μm). Samples were then oven dried (24 h at 60 °C) and weighed to determine the percentage contribution of each fraction. Sediment collected in traps was dried and weighed to give a measure of sediment deposition at each site.

2.6. Contrasting morphological and molecular tools

Analyses of the effects of anthropogenic modification included two factors: Estuary (Es) and Site (Si). Estuary was treated as a fixed factor and site was random with six sites nested within each estuary. All analyses were done in PRIMER v6 with PERMANOVA+ (Anderson, 2001).

Differences in the community composition analysed with morphological and molecular techniques among estuaries were investigated with permutational multivariate analyses of variance (perMANOVA). Data were presence/absence transformed and analyses done on Bray–Curtis similarity matrices (Bray and Curtis, 1957). Homogeneity of dispersion between groups was tested using PERMDISP. Principle components analysis (PCO) was used to visualise differences in sediment communities between estuaries.

To compare the applicability of morphological and molecular tools to estuarine health assessment, biological data sets were analysed with multivariate data sets of predictor variables collected from benthic sediments using distance-based linear modelling (DistLM). Biological data sets were presence/absence transformed and initially analysed against the sediment predictor variables with R^2 selection criteria using forward selection procedure. This identified six variables that explained a significant amount of variation in the morphological data set and 12 variables that explained a significant amount of variation in the molecular data set. The reduced predictor variable data set was then re-analysed with the biological data sets using AIC selection criteria and all specified selection procedure. Results were visualised with distance-based redundancy analysis (dbRDA).

Univariate diversity measures were analysed with ANOVA. Diversity measures included taxa and OTU richness for the entire morphological and molecular data sets. Each data set was also analysed to give a measure of polychaete family richness, crustacean order richness values for each of these taxa. Data were square root transformed and analyses done on Euclidean distance matrices.

Two separate predictor data sets were used when interpreting the potential influence of two major classes of contaminant on the ecological observations. These used either the individual concentrations or contaminant hazard quotients calculated separately for metals (mean sediment quality guideline quotient; metal mSQGQ) and PAHs (PAH mSQGQ), as described previously (Edge et al., 2014; Long, 2006).

3. RESULTS

3.1. Morphological and molecular community composition

Sediment community composition differed significantly between estuaries for both morphological and molecular data sets when taxon identities were matched at a reduced taxonomic resolution (Fig. 1.3; Table 1.1). The taxa that correlated most strongly to differences between sites and estuaries included capitellid, syllid and spionid polychaetes, bivalves and copepods in the morphological data set (Fig. 1.3A). The presence of spionid polychaetes was also strongly correlated with the differences among sites and estuaries in the molecular data set, but other correlated taxa were different from the morphological data set and included terebellid, cirratulid and cossurid polychaetes, gastropods and nemerteans (Fig. 1.3B).

Sediment community composition also differed significantly between estuaries for both morphological and molecular data sets when the highest taxonomic resolution and full data sets were used (Fig. 1.4; Table 1.2).

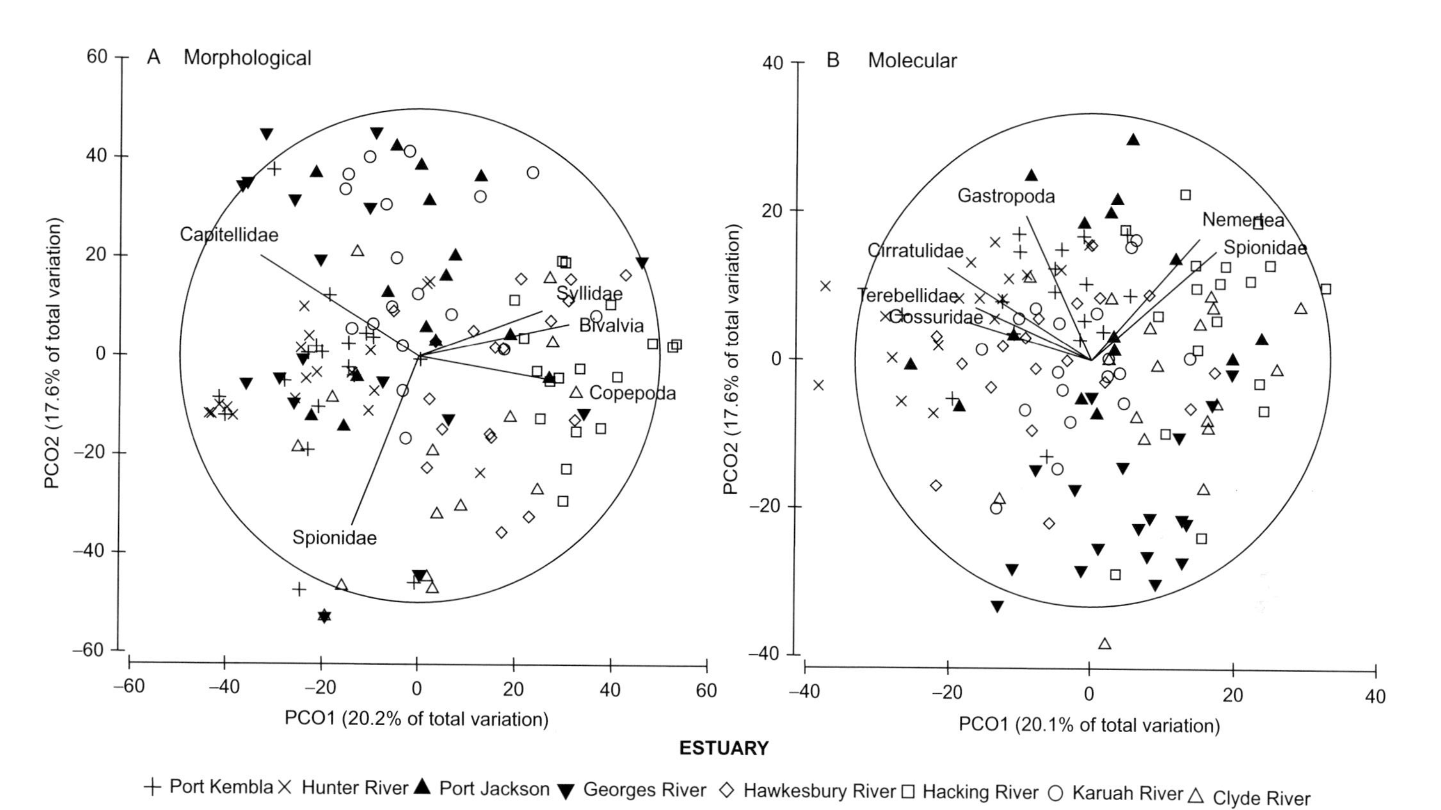

Figure 1.3 PCO plots of the sediment community composition sampled and analysed with (A) morphological and (B) molecular tools. Data sets from each tool were matched to include only taxa sampled in both and were analysed at the same taxonomic level for comparison.

Table 1.1 Permutational multivariate analysis of variance of sediment community composition sampled with (a) morphological and (b) molecular tools
Community composition

Source	df	SS	MS	Pseudo-F	P (perm)
(a) Morphological					
Estuary	7	1.17×10^5	16,696	6.3212	0.0001
Site (Es)	40	1.06×10^5	2645.3	1.6103	0.0001
Res	93	1.53×10^5	1642.8		
(b) Molecular					
Estuary	7	4.87×10^5	6963.8	6.1514	0.0001
Site (Es)	40	4.54×10^5	1134.7	2.2829	0.0001
Res	93	4.62×10^5	497.02		

Data sets from each tool were matched to include only taxa sampled in both and were analysed at the same taxonomic level for comparison.

Differences among estuaries sampled using morphological tools were mostly related to macrofaunal distributions (Fig. 1.4A). In contrast, differences among estuaries sampled using molecular tools appeared to be related most strongly to the presence or absence of different microphytobenthos and meiofauna (Fig. 1.4B). The community sampled from sites in Port Kembla and Hunter River using molecular tools clustered more closely than morphological sampling (Fig. 1.4) and was related to the presence of diatoms, dinoflagellates, ciliphorans and cercozoans as well as the macrofauna nematodes and cirratulid polychaetes (Fig. 1.4B).

Matching taxonomic identities and taxonomic resolution between morphological and molecular data sets resulted in poorly clustered sites within estuaries (Fig. 1.3; Table 1.3). When sediment communities were analysed at the highest taxonomic resolution and included all identifiable taxa, we observed differences in the clustering of sites within estuaries (Fig. 1.4; Table 1.4). Specifically, the average distance among centroids for sites within estuaries were lower in the molecular data set compared to the morphological data set ($t_{df=7}=4.91$, $p<0.01$). Furthermore, using the highest taxonomic resolution increased the clustering of sites within estuaries (decreased the average distances among centroids) in the molecular data set ($t_{df=7}=8.03$, $p<0.01$), but not in the morphological data set ($t_{df=7}=-1.83$, $p>0.05$) (Tables 1.3 and 1.4). This suggests a higher degree

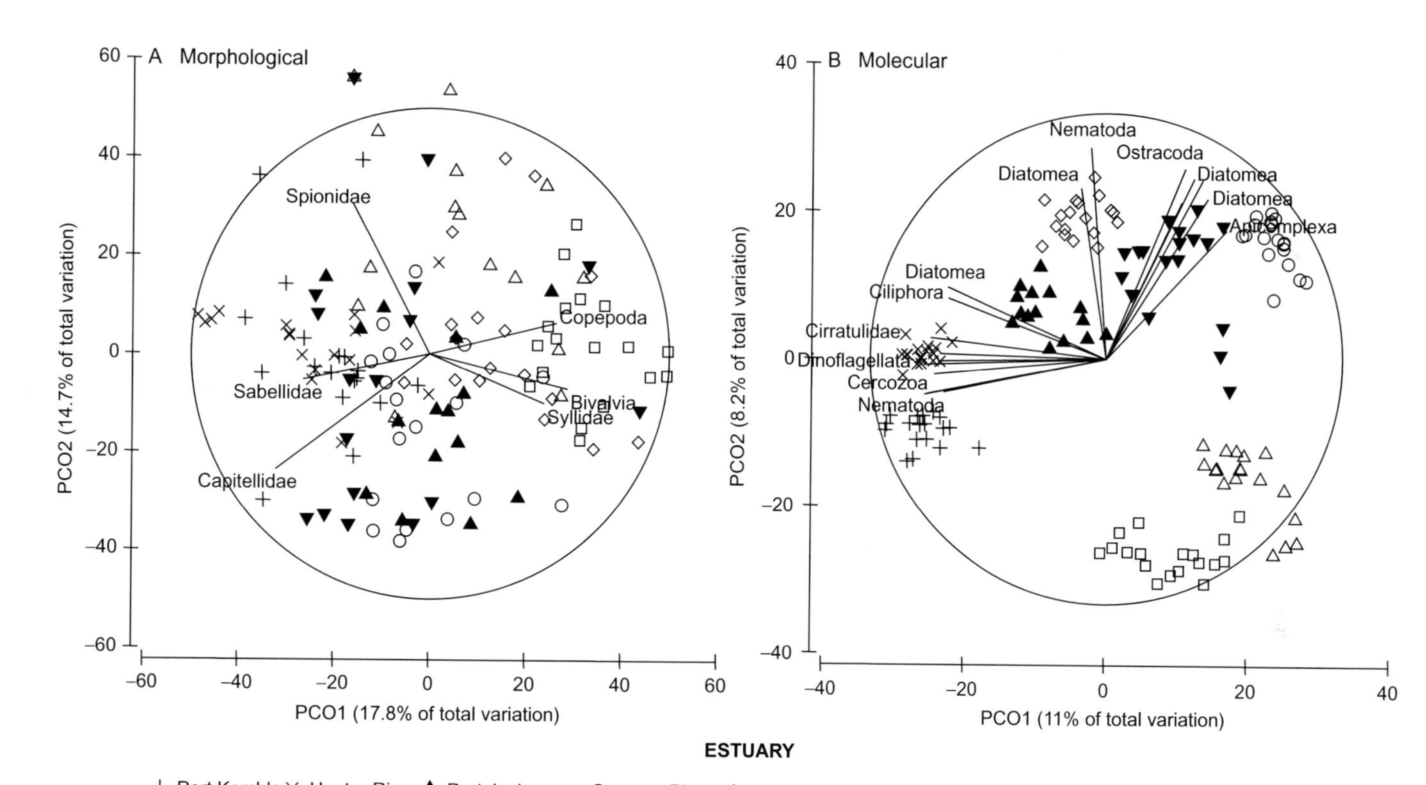

Figure 1.4 PCO plots of the sediment community composition sampled and analysed with (A) morphological and (B) molecular tools. Data sets from each tool included all taxa and were analysed at the highest taxonomic resolution which was family level (morphological) or species level (molecular).

Table 1.2 Permutational multivariate analysis of variance of sediment community composition sampled with (a) morphological and (b) molecular tools
Community composition

Source	df	SS	MS	Pseudo-F	P (perm)
(a) Morphological					
Estuary	7	118,040	16,862	5.90	0.0001
Site (Es)	40	114,500	2863	1.70	0.0001
Res	93	156,840	1687		
(b) Molecular					
Estuary	7	159,740	22,819	7.81	0.0001
Site (Es)	40	117,010	2925	1.74	0.0001
Res	93	156,190	1680		

Data sets from each tool were included all taxa identified by each tool and were analysed at the highest taxonomic resolution. This was family level (morphological) or species level (molecular).

Table 1.3 Average distances among centroids (sites within estuaries) for sediment community composition sampled with (a) morphological and (b) molecular tools
Distance among centroids

Estuary	(a) Morphological	(b) Molecular
Port Kembla	50.13	28.61
Hunter River	54.45	34.40
Port Jackson	26.38	27.36
Georges River	35.07	28.35
Hawkesbury River	30.54	15.44
Hacking River	38.94	27.53
Karuah River	51.11	29.72
Clyde River	33.92	20.11

Data sets from each tool were matched to include only taxa sampled in both and were analysed at the same taxonomic level for comparison.

of similarity in the community composition among sites within estuaries than among sites in different estuaries. Together these results indicate a higher degree of spatial discrimination in the molecular data set than the morphological data set when the highest taxonomic resolution is used and all identifiable organisms are included.

Table 1.4 Average distances among centroids (sites within estuaries) for sediment community composition sampled with (a) morphological and (b) molecular tools
Distance among centroids

Estuary	(a) Morphological	(b) Molecular
Port Kembla	52.91	14.37
Hunter River	56.41	18.17
Port Jackson	42.22	16.00
Georges River	37.81	13.85
Hawkesbury River	31.18	11.97
Hacking River	44.30	16.18
Karuah River	51.48	14.51
Clyde River	32.16	11.27

Data sets from each tool were included all taxa identified by each tool and were analysed at the highest taxonomic resolution. This was family level (morphological) or species level (molecular).

3.2. Relating anthropogenic stressors to sediment communities

Differences in morphological and molecular data sets were fitted to individual metal, PAH and sediment quality predictor variables collected from the benthic sediments (Fig. 1.5; Appendices A–C). Estuaries and sites separated out along two apparent axes with the communities from the more modified estuaries (Port Kembla, Hunter River, Port Jackson and Georges River) that had much greater concentrations of contaminants (metals and/or PAHs) along axis dbRDA1. Sediment quality (silt content and sediment deposition) best predicted the community composition from the relatively unmodified estuaries (Hacking, Karuah, Hawkesbury and Clyde Rivers), which were separated along axis dbRDA2 (Fig. 1.5). Silt content and sediment deposition were particularly good at discriminating among the communities from relatively unmodified estuaries that were sampled with molecular tools (Fig. 1.5B).

The individual metal and PAH predictor variables explained more variation in the molecular data set than the morphological data set (78% compared to 30%; Table 1.5). Metals explained the most variation for both data sets, but this was greater for the molecular compared to the morphological data set and included more individual metals (Table 1.5). The individual PAHs acenapthalene, benzo(a)anthracene and chrysene rather than the

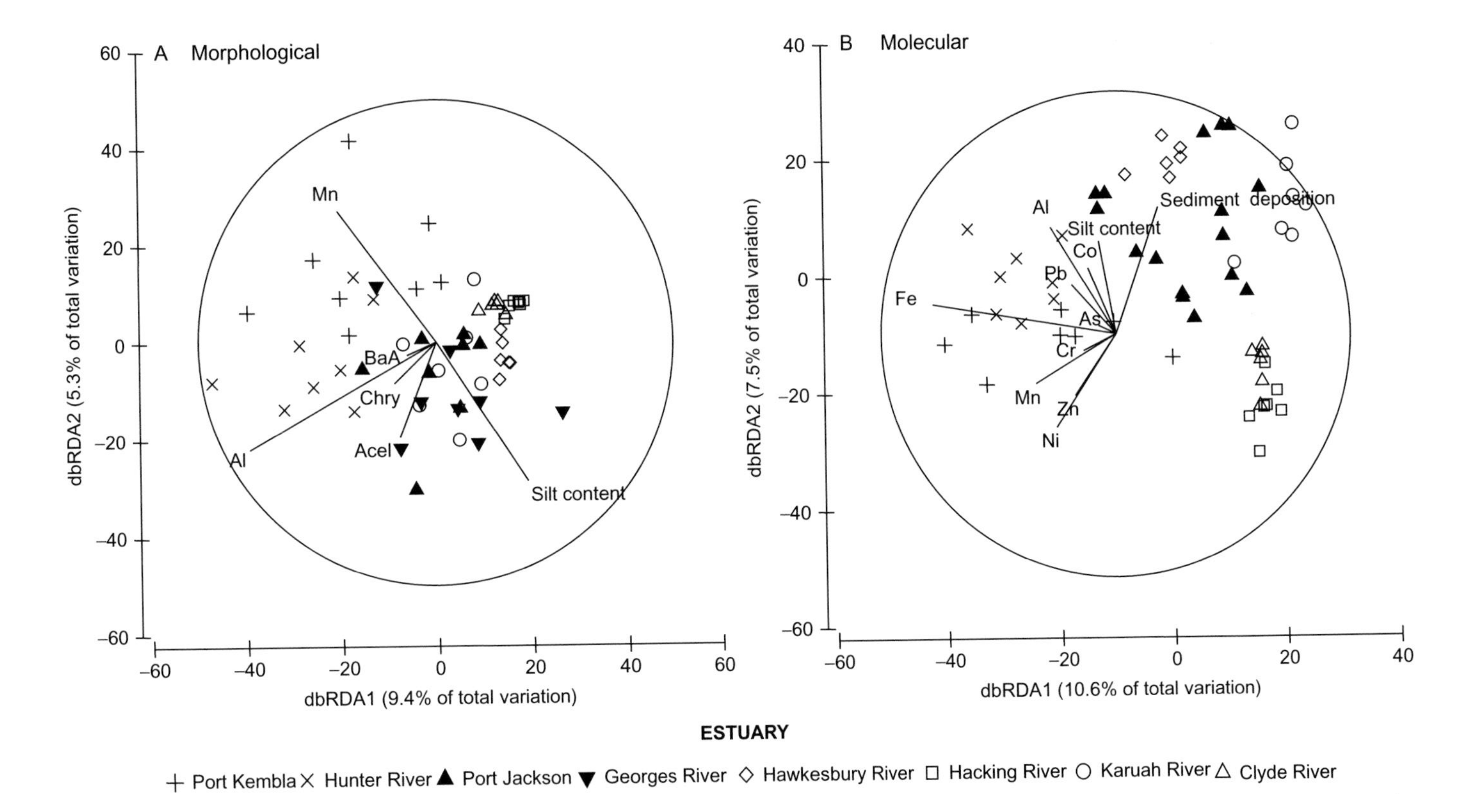

Figure 1.5 Constrained dbRDA plot of sediment communities fitted to significant predictor variables (individual contaminants, organic enrichment indicators and sediment quality) identified using all specified DistLM selection procedure followed by AIC selection criteria (Table 1.5). Lengths of vector overlays indicate the relative influences of fitted predictor variables.

Table 1.5 Proportion of variance in sediment communities sampled with (a) morphological and (b) molecular tools explained by predictor variables including toxic contaminants (individual metals and PAHs), organic enrichment indicators (total organic carbon and chlorophyll a) and measures of sediment quality (silt content and sediment deposition in DistLM marginal tests

Individual contaminant predictors

Variable	SS(trace)	Pseudo-F	P	Proportion
(a) Morphological				
Aluminium	11,725	4.4277	0.0001	0.07
Manganese	8972.3	3.3264	0.0008	0.06
Acenapthalene	7771.1	2.8583	0.0022	0.05
Benzo(a)anthracene	6261.5	2.2805	0.0192	0.04
Chrysene	5626.6	2.0408	0.0474	0.04
Silt content	7673.4	2.8206	0.0029	0.05
Total				0.30
(b) Molecular				
Aluminium	12,248	4.1397	0.0001	0.07
Arsenic	9269.2	3.0777	0.0001	0.05
Cobalt	14,614	5.011	0.0001	0.08
Chromium	14,471	4.9579	0.0001	0.08
Iron	14,491	4.9651	0.0001	0.08
Manganese	14,686	5.0381	0.0001	0.08
Nickel	15,156	5.2143	0.0001	0.09
Lead	11,369	3.8225	0.0001	0.06
Zinc	11,961	4.0359	0.0001	0.07
Sediment deposition	8724.3	2.8874	0.0001	0.05
Silt content	11,696	3.9403	0.0001	0.07
Total				0.78

Predictor variables were identified using all specified DistLM selection procedure followed by AIC selection criteria.

sum of all PAHs (total-PAHs) were also significant predictor variables for the morphological data set, but did not improve the model significantly for the molecular data set (Table 1.5).

When the major contaminant classes (metals and PAHs) were described using SQGQ that represent the mean quotients of the measured concentrations divided by the corresponding sediment quality guideline values (Edge et al., 2014; Long, 2006), similar but weaker patterns were observed compared to when individual variables were included in the models (Fig. 1.6; Table 1.6). Estuaries and sites again separated out along two apparent axes with the communities from the more modified estuaries related to contaminants (metals and PAHs) and also organic enrichment indicators (TOC and chlorophyll a) along axis dbRDA1 (Fig. 1.6). Sediment quality (silt content and sediment deposition) best predicted the community composition from the relatively unmodified estuaries, which were separated along axis dbRDA2 (Fig. 1.6). The gradient of silt content revealed two distinct groupings of relatively unmodified estuaries with the communities in Hawkesbury and Karuah Rivers predicted by higher silt content and the communities in Hacking and Clyde Rivers predicted by coarser sediments (Fig. 1.6).

The use of SQGQs for the major contaminant classes reduced the predictive ability compared to using individual contaminant concentrations for both morphological and molecular data sets. The magnitude of explanatory power lost was 45% and 65% for the molecular and morphological data sets, respectively. However, whether individual metals or the mean SQGQ for metals was used, metal contaminants explained the most variation in both the morphological and molecular data sets (19% and 35%, respectively; Table 1.6). Silt content, chlorophyll-a concentrations and TOC content also significantly increased the explanatory power of the model for both the morphological and molecular data sets (Table 1.6). Sediment deposition and PAHs as represented by a mean sediment quality guidelines quotient also explained a significant amount of variation in the molecular data set (Table 1.6a), but did not improve the model for the morphological data set (Table 1.6b).

3.3. Diversity measures

The morphological and molecular data sets had similar patterns for some richness measures and varied significantly among estuaries and sites within estuaries (Table 1.7; Fig. 1.7). The exceptions included polychaete family richness

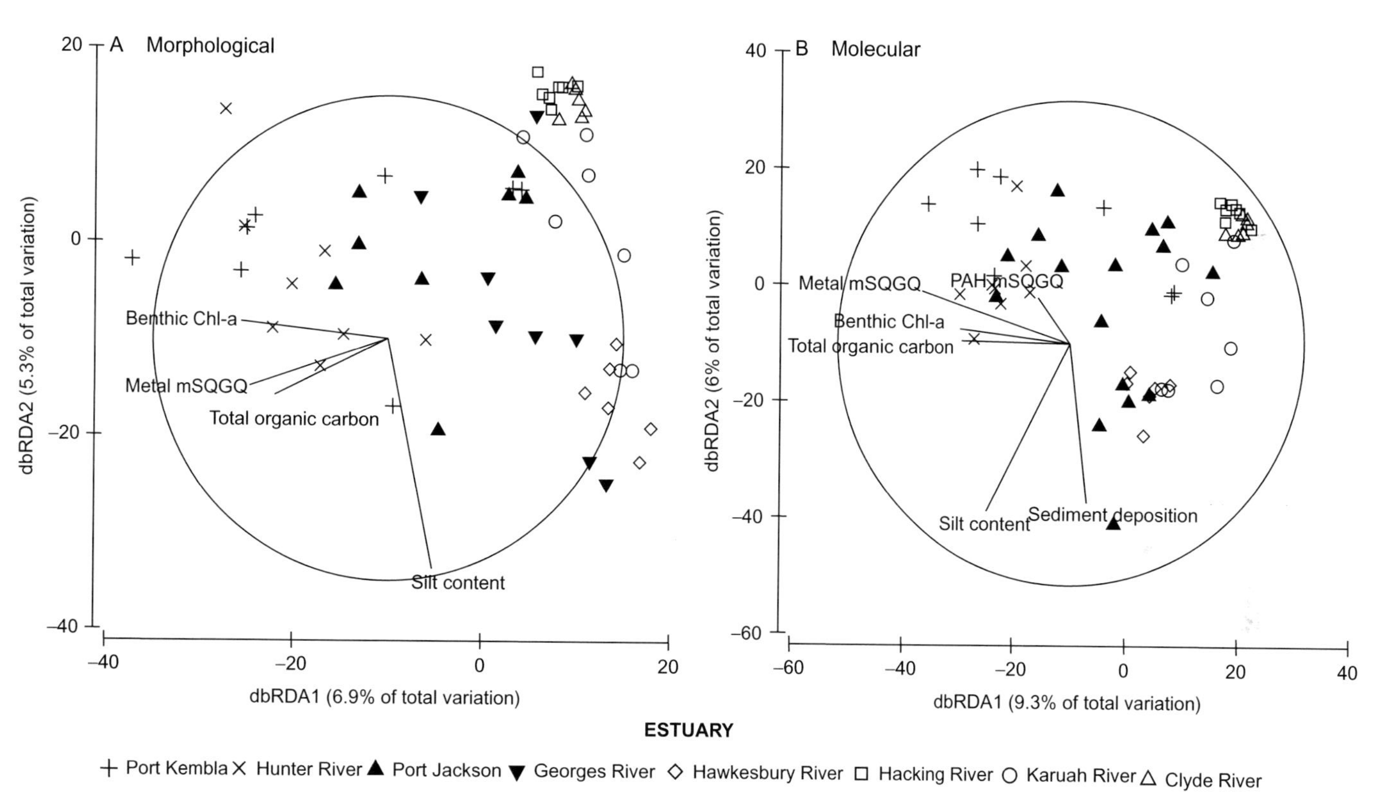

Figure 1.6 Constrained dbRDA plot of sediment communities fitted to significant predictor variables (contaminant quotients, organic enrichment indicators and sediment quality) identified using all specified DistLM selection procedure followed by AIC selection criteria (Table 1.6). Lengths of vector overlays indicate the relative influences of fitted predictor variables.

Table 1.6 Proportion of variance in sediment communities sampled with (a) morphological and (b) molecular tools explained by predictor variables including mean sediment quality guideline quotients (mSQGQs) for metals and PAHs, organic enrichment indicators (total organic carbon and chlorophyll a) and measures of sediment quality (silt content and sediment deposition) in DistLM marginal tests
Contaminant quotient predictors

Variable	SS(trace)	Pseudo-F	P	Prop.
(a) Morphological				
Metal mSQGQ	8191.8	3.0214	0.0018	0.05
Silt content	7673.4	2.8206	0.0043	0.05
Chlorophyll a	7174.9	2.6288	0.0043	0.04
Total organic carbon	6894.4	2.5214	0.0071	0.04
Total				0.19
(b) Molecular				
Metal mSQGQ	13,812	4.713	0.0001	0.08
PAH mSQGQ	5179.5	1.679	0.0012	0.03
Sediment deposition	8724.3	2.8874	0.0001	0.05
Silt content	11,696	3.9403	0.0001	0.07
Chlorophyll a	11,536	3.8827	0.0001	0.06
Total organic carbon	10,484	3.5063	0.0001	0.06
Total				0.35

Predictor variables were identified using all specified DistLM selection procedure followed by AIC selection criteria.

measured by molecular tools and crustacean order richness measured by morphological tools (Table 1.7). Both techniques found the highest taxa and OTU richness in Hacking River (Fig. 1.7). Variation in OTU richness was greater within the unmodified estuaries compared to the modified estuaries (Table 1.7; Fig. 1.6). Polychaete family richness was in the same order of magnitude for both the morphological and molecular data sets (<11 families) (Table 1.7; Fig. 1.7). In contrast, the molecular data set detected consistently higher crustacean order richness compared to the morphological data set, ~7 orders in four estuaries for the molecular data compared to a maximum of 6 orders in only one estuary for the morphological data set (Table 1.7; Fig. 1.7).

Table 1.7 Univariate analysis of variance of richness measures sampled with (a) morphological and (b) molecular tools

Source	df	SS	MS	Pseudo-F	P (perm)
Taxa richness					
(a) Morphological					
Estuary	7	44.46	6.35	10.73	0.0001
Site (Es)	40	23.72	0.59	1.76	0.0132
Res	93	31.31	0.34		
(b) Molecular					
Estuary	7	78.75	11.25	4.25	0.0017
Site (Es)	40	106.15	2.65	2.41	0.0004
Res	93	102.60	1.10		
Polychaete family richness					
(a) Morphological					
Estuary	7	29.71	4.24	9.91	0.0001
Site (Es)	40	17.16	0.43	1.64	0.0255
Res	93	24.27	0.26		
(b) Molecular					
Estuary	7	6.93	0.99	1.90	0.0855
Site (Es)	40	20.87	0.52	2.93	0.0002
Res	93	16.58	0.18		
Crustacean order richness					
(a) Morphological					
Estuary	7	28.56	4.08	10.75	0.0001
Site (Es)	40	15.18	0.38	1.14	0.3038
Res	93	31.02	0.33		
(b) Molecular					
Estuary	7	3.46	0.49	4.63	0.0009
Site (Es)	40	4.27	0.11	1.73	0.0167
Res	93	5.74	0.06		

Richness measures included taxa (OTU) richness, polychaete family richness and crustacean order richness.

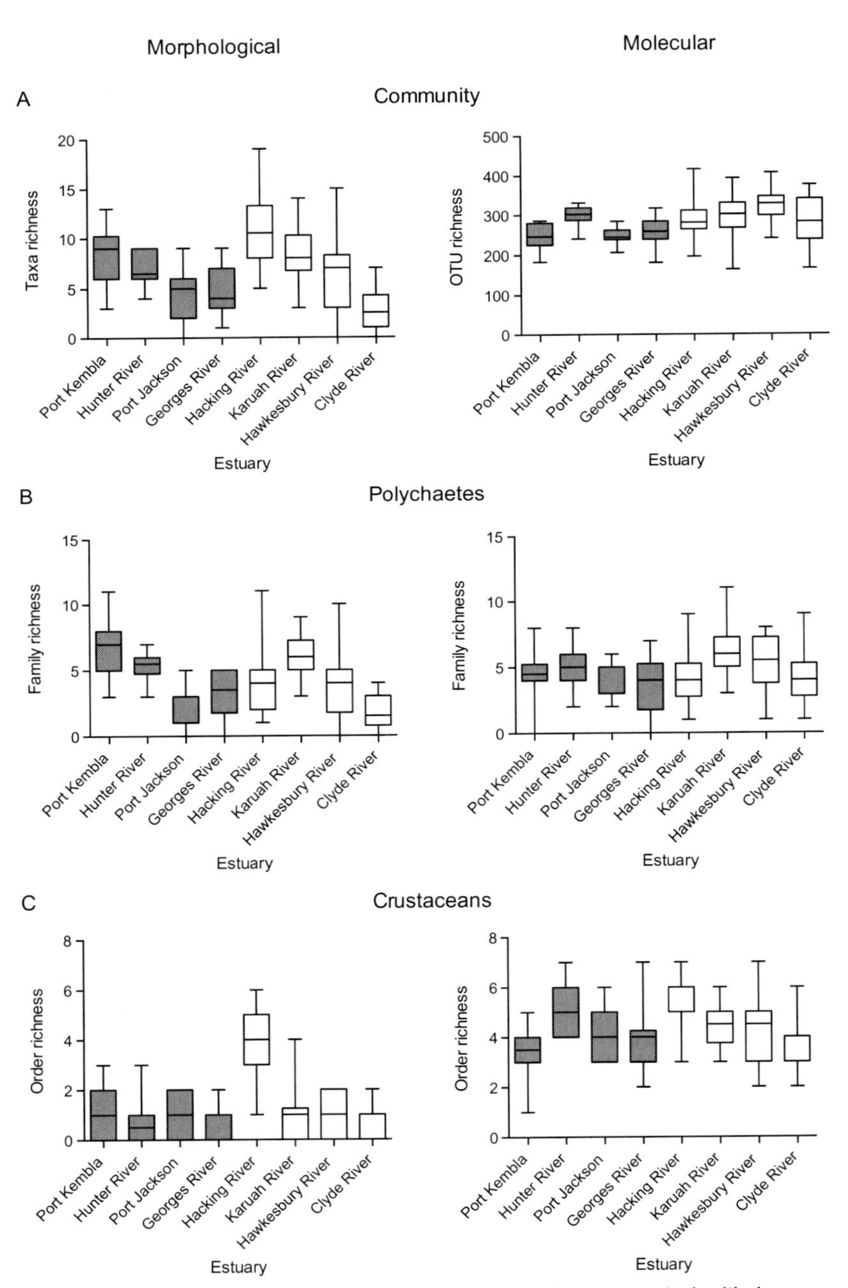

Figure 1.7 Boxplots showing the full range of variation (min to max), the likely range of variation (between third and first quartile) and the median for diversity measures analysed from morphological and molecular data sets. Diversity measures include (A) taxa/OTU richness, (B) polychaete family richness and (C) crustacean order richness.

4. DISCUSSION

Next-generation DNA sequencing platforms allow us to generate big data scale observations of biological assemblages (Baird and Hajibabaei, 2012). We examined the potential advantages of new molecular techniques over traditional morphological tools for assessing ecological change using co-sampled sediments from a large-scale field study of estuary health. When the two ecological observation tools were compared using the same taxonomic level and limited to organisms identified in both data sets, a high degree of spatial variability between sites within estuaries was identified using both approaches. However, the ability to undertake a higher level of taxonomic resolution and including all organisms allowed for increased discrimination among estuaries using the molecular approach. The greater potential to discriminate among sampling regions is useful for biomonitoring purposes where assessments often require region-specific information (Graham et al., 1991). Furthermore, the information provided by molecular techniques appeared to be more sensitive to a range of well-established drivers of benthic ecology (both natural environment variables and metals and PAHs). The morphological approach allowed for greater discrimination between the influence of metals, PAHs and environmental variables when individual variables were combined into quotients representative of the classes of contaminants (metals and PAHs) that represented the major stressors in the system. Apart from higher taxonomic resolution, there was also an order of magnitude more taxonomic units recorded in the molecular approach relative to the morphological when all organisms detected by each method were considered. While the morphological data set was constrained to macroinvertebrate sampling, the molecular tools could be used to sample a wide range of other taxa from the microphytobenthos including, e.g., diatoms and dinoflagellates. A separate comparison of the macrofauna only (polychaetes and crustacea) revealed strikingly similar patterns of diversity at family and order levels. Hence, we conclude that molecular approaches are now sufficiently advanced to provide not only equivalent information to that collected using traditional morphological approaches, but also an order of magnitude bigger, better, faster data collection to more precisely differentiate between environmental samples.

4.1. Characterising ecological systems

Thorough characterisation of ecological systems is essential to inform environmental management (Baird and Hajibabaei, 2012). Furthermore, to ensure generality across regions and systems requires standard methods of observing change (Birk et al., 2012; Friberg et al., 2011). Past methods have been constrained by cost and time, with traditional morphological assessments requiring extensive sampling effort and taxonomic expertise that often needs to be region-specific (Menezes et al., 2010). These assessments investigate different groups of organisms, often macroinvertebrates, at various levels of taxonomic resolution, often family or genus. But the variation in methodology often prohibits comparisons between different studies. To improve generality, more than 300 biotic indices have been developed for comparisons across regions and to answer the question of whether an area is impacted (Birk et al., 2012). Despite this, the quantity of information gained through surveys using traditional morphological approaches is seldom sufficient to form the backbone of environmental management plans.

Effective biomonitoring strategies need to match the scale and quality of biological data with associated stressor information. Past morphological approaches have been limited by taxonomic clarity and processing time, but now next-generation sequencing techniques can provide detailed biological information that is not dependent on traditional taxonomic classifications and can be processed in a relatively short time period. We compared morphological and molecular approaches to characterise estuarine sediment communities using exactly the same field sampling effort and observed greater discrimination between estuaries using the molecular data. Samples clustered tightly within estuaries with almost no overlap in the communities when analysed with molecular techniques. Major estuary indicator groups were microphytobenthos including the pico- and meio-eukaryotic fractions (e.g. fungi, diatoms and dinoflagellate species) that are not measured by traditional macroinvertebrate studies. These fractions are commonly found to act as sensitive indicator groups for anthropogenic stress (Chariton et al., 2010a, 2014; Traunspurger and Drews, 1996).

Spatial variability of sediment communities appeared similar among approaches when they were assessed at the same taxonomic resolution, but increased discrimination of location was clearer from the molecular approach when using the full capacity for taxonomic resolution. Increased discrimination has also been observed in river samples assessed using both morphological and molecular tools (Hajibabaei et al., 2011). The major point of difference between the morphological and molecular techniques was the order of magnitude greater richness observed using the molecular tools. This increased the power of discrimination between samples, and resulted from greater taxonomic clarity and identification beyond just the macroinvertebrates.

In contrast, estuaries were not well separated by morphological sampling techniques with a great deal of overlap between locations. Indicator groups included different polychaete families and crustaceans. These groups are often included in biotic indices due to their observed sensitivity to environmental change (Dauvin, 2007). Polychaetes in particular have opportunistic representative families, e.g., Capitellidae with documented abundance increases in response to organic enrichment (Pearson and Rosenberg, 1978). Polychaete family richness and abundance has been shown to increase in estuaries exposed to increased anthropogenic modification (Dafforn et al., 2013). A comparison of the macrofauna only (polychaetes and crustaceans) revealed strikingly similar patterns of occurrence and diversity at family and order levels. The molecular approach therefore provided significantly more information than morphological samples.

4.2. Distinguishing the effects of multiple stressors

Ecosystems are increasingly exposed to multiple stressors and effective biomonitoring needs to be able to distinguish the effects of different stressors. This presents major challenges for studies on estuarine communities since they are exposed to dramatic changes in both natural environmental stressors and stressors arising from anthropogenic activities (Elliott and Quintino, 2007). The combined effects of multiple stressors are difficult to distinguish in large-scale surveys since many contaminants are highly correlated with each other and with the percent fines in the sediment (Simpson et al., 2005). The biological information collected from the molecular approach had a greater capacity to discriminate between potential drivers from a range of potential variables compared to the biological information from the morphological approach.

Communities sampled from both approaches separated out on two distinct gradients (metals/PAHs and silt content in the sediments). Metals and PAHs were the strongest predictors of sediment community composition regardless of sampling approach. When individual variables were considered they explained more than double the proportion of variance in the molecular community compared to the morphological community. The two heavily industrialised estuaries, Port Kembla and Newcastle Harbour, and, to a lesser extent, the highly urbanised Port Jackson, spread out along the contaminant axis, separate from the less modified estuaries. Laboratory and field studies have found changes in sediment community composition to occur with increasing metal concentrations. Sediments contaminated with 30 mg/kg Cu were associated with reduced infaunal diversity (Stark et al., 2003) and >300 mg/kg Cu can reduce recolonisation success of infauna (Rygg, 1985). Sediment metal contaminant concentrations in the current study were above guideline values predicted to have ecological effects (up to 1250 mg/kg Cu, 550 mg/kg Pb and 900 mg/kg Zn) and these may have contributed to the shifts we observed in community composition in the more industrialised estuaries. Shifts in bacterial community composition have also been observed in association with contamination in similar sediments (Sun et al., 2013). In addition, discrimination of estuaries was also possible based on silt content. Grain size is recognised as a major driver of sediment assemblages (Dafforn et al., 2012, 2013; Sun et al., 2012).

4.3. The 'new diversity'

Most management approaches aim to protect biological diversity at a species level. However, until recently species level information has too often been unavailable due to the inherent difficulties of identification and the expensive nature of accurate taxonomic work. Often proxies are used, such as indicator species or habitat formers (Hilty and Merenlender, 2000), or single species assessments using toxicity tests (Greenstein et al., 2008; Simpson and Spadaro, 2011). However, the ultimate aim is to preserve local, regional and global diversity patterns to support ecosystem function and stability. Proxies therefore have their weaknesses if they are not appropriate representatives of the wider ecosystem. In particular, habitat forming species or physical proxies such as substrate/soil/altitude might vary over large spatial scales while the associated fauna vary on smaller scales (Wiens, 1989). It is therefore ideal if biomonitoring approaches target diversity at the species level. This will provide detailed patterns of alpha, beta and gamma diversity that are important to detect and monitor for conservation purposes, range changes and biological invasion.

Post-processing in preparation for morphological identification involves the selection of a particular size-class of organisms through sieving and then a live stain excludes any non-living material. Therefore, morphological approaches sample what is currently living at that particular site and time. In contrast, molecular analyses provide information on organisms living and dead from a large variety of size classes. Since samples are collected from surficial sediments, they will also contain a pelagic signature from detrital rain. These factors combined contribute to the increased information available from molecular techniques that can allow us to distinguish responses over longer time periods and greater spatial scales as well as differentiating between multiple stressors. While the potential for molecular approaches to sample living and dead organisms creates a useful time-integrated response to stressors, this data may provide challenges when interpreting a recent impact as any lethal effects may be masked by a molecular approach. However, it is also possible to retrieve only the signature of 'active' organisms using RNA approaches rather than DNA.

It must also be recognised that the 'new diversity' revealed through molecular techniques is distinct from the taxonomic classification implemented by traditional morphological sampling. DNA sequences cannot be directly translated into species level information. In fact the OTUs that are assigned to different sequences can represent the same taxa that have different evolutionary histories. This is often referred to as 'cryptic diversity' because it is hidden beneath that observable from gross morphology or current phenotype (or a result of divergence via traits we do not measure as phenotype). It may be viewed as being "ecotype" diversity and can be critical in maintaining species at times of environmental change.

Post-sampling strategies for molecular data to assign taxonomic information to sequences are therefore diverse and attempts to standardise processing and analyses are challenging due to the rapid progression of understanding in this field. A range of processing software packages exists to remove duplicate sequences introduced during the PCR amplification process, to cluster sequences based on similarity and more accurately estimate phylogenetic composition. Denoising and chimaera removal are additional steps to correct problems in sequence data and improve biological accuracy. While these different approaches to preparing the raw sequences will impact diversity estimates to some extent, the overall trends remain the same (Morgan et al., 2013). Furthermore, many of the problems that exist for processing molecular data are somewhat ephemeral as platforms evolve rapidly. This ephemerality points to other drawbacks to using rapidly developing technologies, the potential difficulty in comparing across studies and the potential of longer-term redundancy. When different studies utilise alternative marker genes or primer sets to generate amplicons, meta-analyses must be carried out based on a either existing taxonomy (likely of a fairly low resolution), or by potentially time consuming reanalysis of data against a curated database containing all required elements (e.g. Full-length 18S rRNA gene sequences that cover different variable regions used in different studies). However, such databases are rapidly being generated (e.g. SILVA) and the computational requirements for reanalysis being continually improved. Furthermore, sequencing technologies are trending towards longer reads that will simplify comparisons with shorter reads using methods such as fragments recruitment.

Biomonitoring aims to characterise ecological systems and detect change in response to environmental and anthropogenic stressors. Molecular techniques allow for higher taxonomic resolution and generate information that is potentially more sensitive to ecological change because of the whole ecosystem comparisons that are possible.

4.4. Conclusion

The molecular approaches used in this study are sufficiently advanced to provide not only the information collected using traditional morphological approaches, but also an order of magnitude bigger, better, faster data to differentiate locations, and to investigate cause in complex stressor scenarios. Communities analysed through metagenomics analysis were more clearly characterised in terms of their taxonomic complement, yielding clearer responses to environmental stressors and better spatial discrimination at a scale critical for biomonitoring. In addition to permitting more rapid sample processing at lower cost, the higher taxonomic resolution offered by the DNA approach is perhaps the greatest prize offered by metagenomic analysis of environmental samples. However, assigning taxonomic meaning to bulk sequence data is currently constrained by the availability of reliable, reproducible sequences obtained from taxonomically verified specimens. While techniques are being developed to assign taxonomic rank using algorithmic approaches based on sequence homology (e.g. Porter et al., 2014), there is still a need to continue to provide verified specimen sequences to libraries. However, given that so much of the world's flora and fauna may not be currently known in either taxonomic or molecular terms (Adams et al., 2014), use of both DNA barcoding and molecular classifiers is encouraged, as no single method can reliably detect everything in a single sample.

ACKNOWLEDGEMENTS

This research was primarily supported by the Australian Research Council through an Australian Research Fellowship (DP1096900) awarded to Johnston and a Linkage Grant (LP0990640) awarded to Johnston, Kelaher and Coleman. We would like to thank members of the Subtidal Ecology and Ecotoxicology laboratory and volunteers for their assistance. We would also like to thank Bluescope Steel and the NSW Marine Parks Authority for their support. In addition, we wish to thank CSIRO's Oceans and Atmosphere Flagship and CSIRO's OCE Distinguished Visiting Scientists Program. D. J. B. would also like to acknowledge support through the Canadian Natural Sciences and Engineering Research Council's Discovery Grant Program.

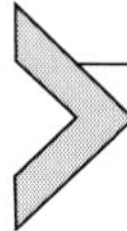

APPENDIX A. METAL CONTAMINANT CONCENTRATIONS (mg/kg DRY WT) IN BENTHIC SEDIMENTS

Estuary		As	Cd	Co	Cr	Cu	Ni	Pb	Zn
Port Kembla	Mean	13	1	9	65	149	16	141	678
	Min.	8	1	4	31	63	9	64	266
	Max.	16	2	14	91	224	26	245	1494
Hunter River	Mean	10	2	15	40	75	34	122	491
	Min.	10	1	11	30	52	25	49	252
	Max.	11	3	17	47	91	43	234	918
Port Jackson	Mean	20	1	7	68	113	10	189	433
	Min.	14	0	4	31	63	5	112	238
	Max.	25	1	9	94	166	14	293	586
Georges River	Mean	16	1	7	27	33	10	72	192
	Min.	4	0	3	12	26	6	55	126
	Max.	23	1	11	37	42	15	102	238
Hacking River	Mean	8	0	1	5	5	1	25	80
	Min.	7	0	1	3	4	1	13	21
	Max.	10	0	2	6	8	2	48	286
Karuah River	Mean	7	0	4	7	3	4	10	34
	Min.	6	0	3	5	1	2	8	25
	Max.	12	0	6	11	6	6	13	45
Hawkesbury River	Mean	8	0	1	5	5	1	25	80
	Min.	7	0	1	3	4	1	13	21
	Max.	10	0	2	6	8	2	48	286
Clyde River	Mean	6	0	4	5	1	5	5	22
	Min.	3	0	2	4	0	4	4	16
	Max.	12	0	5	9	3	7	8	34

APPENDIX B. PRIORITY PAH CONTAMINANT CONCENTRATIONS (μg/kg DRY WT) IN BENTHIC SEDIMENTS

Estuary		Nap	Acel	Ace	Fl	PhA	An	FlA	Py	BaA	Chry	BbfF	BaP	IP	DahA	BghiP	Total-PAHs
Port Kembla	Mean	11	1	0	1	4	1	4	4	2	2	3	2	1	0	2	37
	Min.	2	0	0	0	1	0	1	1	0	1	1	1	0	0	0	9
	Max.	20	1	1	1	6	3	8	7	4	3	6	4	2	1	3	66
Hunter River	Mean	0	0	0	0	0	0	1	1	0	0	1	0	0	0	0	5
	Min.	0	0	0	0	0	0	1	1	0	0	0	0	0	0	0	3
	Max.	0	0	0	0	1	0	2	2	1	1	1	1	0	0	0	9
Port Jackson	Mean	0	0	0	0	1	0	3	3	2	1	3	3	1	0	1	21
	Min.	0	0	0	0	0	0	1	1	0	0	0	1	0	0	0	5
	Max.	0	0	0	0	4	1	7	7	3	3	6	5	2	1	2	42
Georges River	Mean	0	0	0	0	0	0	0	0	0	0	0	0	0	0	0	2
	Min.	0	0	0	0	0	0	0	0	0	0	0	0	0	0	0	0
	Max.	0	0	0	0	0	0	0	0	0	0	1	0	0	0	0	3

Hacking River	Mean	0	0	0	0	0	0	0	0	0	0	0	0	0	0	0	0
	Min.	0	0	0	0	0	0	0	0	0	0	0	0	0	0	0	0
	Max.	0	0	0	0	0	0	0	0	0	0	0	0	0	0	0	0
Karuah River	Mean	0	0	0	0	0	0	0	0	0	0	0	0	0	0	0	2
	Min.	0	0	0	0	0	0	0	0	0	0	0	0	0	0	0	0
	Max.	0	0	0	0	0	0	1	1	0	0	0	0	0	0	0	3
Hawkesbury River	Mean	0	0	0	0	0	0	0	0	0	0	0	0	0	0	0	0
	Min.	0	0	0	0	0	0	0	0	0	0	0	0	0	0	0	0
	Max.	0	0	0	0	0	0	0	0	0	0	0	0	0	0	0	0
Clyde River	Mean	0	0	0	0	0	0	0	0	0	0	0	0	0	0	0	0
	Min.	0	0	0	0	0	0	0	0	0	0	0	0	0	0	0	0
	Max.	0	0	0	0	0	0	0	0	0	0	0	0	0	0	0	1

APPENDIX C. SEDIMENT QUALITY (SILT CONTENT (% <63 μm)) AND ENRICHMENT MEASURES (CHLOROPHYLL A (μg/g) AND TOTAL ORGANIC CARBON (%)) IN BENTHIC SEDIMENTS

Estuary		Silt content	Sediment deposition	Chlorophyll a	Total organic carbon
Port Kembla	Mean	23	49	2	7
	Min.	4	5	0	1
	Max.	52	140	3	13
Hunter River	Mean	60	49	16	2
	Min.	18	39	5	2
	Max.	92	64	22	3
Port Jackson	Mean	39	31	6	2
	Min.	13	17	1	1
	Max.	68	36	11	3
Georges River	Mean	52	176	4	2
	Min.	1	51	0	0
	Max.	88	299	8	3
Hacking River	Mean	1	6	1	0
	Min.	0	2	0	0
	Max.	5	15	2	1
Karuah River	Mean	36	126	1	1
	Min.	11	22	0	0
	Max.	65	191	3	2
Hawkesbury River	Mean	1	112	1	0
	Min.	0	48	0	0
	Max.	5	185	2	1
Clyde River	Mean	4	24	0	0
	Min.	0	4	0	0
	Max.	8	29	0	1

REFERENCES

Adams, M., Raadik, T.A., Burridge, C.P., Georges, A., 2014. Global biodiversity assessment and hyper-cryptic species complexes: more than one species of elephant in the room? Syst. Biol. 63, 518–533.

Akin, S., Winemiller, K.O., Gelwick, F.P., 2003. Seasonal and spatial variations in fish and macrocrustacean assemblage structure in Mad island marsh estuary, Texas. Estuar. Coast. Shelf Sci. 57, 269–282.

Anderson, M.J., 2001. A new method for non-parametric multivariate analysis of variance. Austral Ecol. 26, 32–46.

ANZECC/ARMCANZ, 2000. Australian and New Zealand guidelines for freshwater and marine water quality. Australian and New Zealand Environment and Conservation Council and Agriculture and Resource Management of Australia and New Zealand, Canberra.

Baird, D.J., Hajibabaei, M., 2012. Biomonitoring 2.0: a new paradigm in ecosystem assessment made possible by next-generation DNA sequencing. Mol. Ecol. 21, 2039–2044.

Birk, S., Bonne, W., Borja, A., Brucet, S., Courrat, A., Poikane, S., Solimini, A., van de Bund, W., Zampoukas, N., Hering, D., 2012. Three hundred ways to assess Europe's surface waters: an almost complete overview of biological methods to implement the water framework directive. Ecol. Indic. 18, 31–41.

Birk, S., Willby, N.J., Kelly, M.G., Bonne, W., Borja, A., Poikane, S., van de Bund, W., 2013. Intercalibrating classifications of ecological status: Europe's quest for common management objectives for aquatic ecosystems. Sci. Total Environ. 454–455, 490–499.

Bonada, N., Prat, N., Resh, V.H., Statzner, B., 2006. Developments in aquatic insect biomonitoring: a comparative analysis of recent approaches. Annu. Rev. Entomol. 51, 495–523.

Borja, A., Bricker, S.B., Dauer, D.M., Demetriades, N.T., Ferreira, J.G., Forbes, A.T., Hutchings, P., Jia, X., Kenchington, R., Marques, J.C., Zhu, C., 2008. Overview of integrative tools and methods in assessing ecological integrity in estuarine and coastal systems worldwide. Mar. Pollut. Bull. 56, 1519–1537.

Bray, J.R., Curtis, J.T., 1957. An ordination of the upland forest communities of southern Wisconsin. Ecol. Monogr. 27, 325–349.

Brown, M.V., Philip, G.K., Bunge, J.A., Smith, M.C., Bissett, A., Lauro, F.M., Fuhrman, J.A., Donachie, S.P., 2009. Microbial community structure in the north pacific ocean. ISME J. 3, 1374–1386.

Burton, G.A., Johnston, E.L., 2010. Assessing contaminated sediments in the context of multiple stressors. Environ. Toxicol. Chem. 29, 2625–2643.

Cao, Y., Hawkins, C.P., 2011. The comparability of bioassessments: a review of conceptual and methodological issues1. J. N. Am. Benthol. Soc. 30, 680–701.

Caporaso, J.G., Kuczynski, J., Stombaugh, J., Bittinger, K., Bushman, F.D., Costello, E.K., Fierer, N., Pena, A.G., Goodrich, J.K., Gordon, J.I., 2010. QIIME allows analysis of high-throughput community sequencing data. Nat. Methods 7, 335–336.

Chapman, P.M., Wang, F., 2001. Assessing sediment contamination in estuaries. Environ. Toxicol. Chem. 20, 3–22.

Chariton, A.A., Court, L.N., Hartley, D.M., Colloff, M.J., Hardy, C.M., 2010a. Ecological assessment of estuarine sediments by pyrosequencing eukaryotic ribosomal DNA. Front. Ecol. Environ. 8, 233–238.

Chariton, A.A., Roach, A.C., Simpson, S.L., Batley, G.E., 2010b. Influence of the choice of physical and chemistry variables on interpreting patterns of sediment contaminants and their relationships with estuarine macrobenthic communities. Mar. Freshw. Res. 61, 1109–1122.

Chariton, A.A., Ho, K.T., Proestou, D., Bik, H., Simpson, S.L., Portis, L.M., Cantwell, M.G., Baguley, J.G., Burgess, R.M., Pelletier, M.M., Perron, M., Gunsch, C., Matthews, R.A.,

2014. A molecular-based approach for examining responses of eukaryotes in microcosms to contaminant-spiked estuarine sediments. Environ. Toxicol. Chem. 33, 359–369.

Charmantier, G., Haond, C., Lignot, J., Charmantier-Daures, M., 2001. Ecophysiological adaptation to salinity throughout a life cycle: a review in homarid lobsters. J. Exp. Biol. 204, 967–977.

Coissac, E., Riaz, T., Puillandre, N., 2012. Bioinformatic challenges for DNA metabarcoding of plants and animals. Mol. Ecol. 21, 1834–1847.

Creer, S., Fonseca, V.G., Porazinska, D.L., Giblin-Davis, R.M., Sung, W., Power, D.M., Packer, M., Carvalho, G.R., Blaxter, M.L., Lambshead, P.J.D., 2010. Ultrasequencing of the meiofaunal biosphere: practice, pitfalls and promises. Mol. Ecol. 19, 4–20.

Dafforn, K.A., Simpson, S.L., Kelaher, B.P., Clark, G.F., Komyakova, V., Wong, C.K.C., Johnston, E.L., 2012. The challenge of choosing environmental indicators of anthropogenic impacts in estuaries. Environ. Pollut. 163, 207–217.

Dafforn, K.A., Kelaher, B.P., Simpson, S.L., Coleman, M.A., Hutchings, P.A., Clark, G.F., Knott, N.A., Doblin, M.A., Johnston, E.L., 2013. Polychaete richness and abundance enhanced in anthropogenically modified estuaries despite high concentrations of toxic contaminants. PLoS One 8, e77018.

Dauvin, J.-C., 2007. Paradox of estuarine quality: benthic indicators and indices, consensus or debate for the future. Mar. Pollut. Bull. 55, 271–281.

Edge, K.J., Dafforn, K.A., Simpson, S.L., Roach, A.C., Johnston, E.L., 2014. A biomarker of contaminant exposure is effective in large scale assessment of ten estuaries. Chemosphere 100, 16–26.

Elliott, M., Quintino, V., 2007. The estuarine quality paradox, environmental homeostasis and the difficulty of detecting anthropogenic stress in naturally stressed areas. Mar. Pollut. Bull. 54, 640–645.

Elliott, M., Cutts, N.D., Trono, A., 2014. A typology of marine and estuarine hazards and risks as vectors of change: a review for vulnerable coasts and their management. Ocean Coast. Manag. 93, 88–99.

Friberg, N., Bonada, N., Bradley, D.C., Dunbar, M.J., Edwards, F.K., Grey, J., Hayes, R.B., Hildrew, A.G., Lamouroux, N., Trimmer, M., 2011. Biomonitoring of human impacts in freshwater ecosystems: the good, the bad and the ugly. Adv. Ecol. Res. 44, 1–68.

Gardham, S., Hose, G.C., Stephenson, S., Chariton, A.A., 2014. DNA metabarcoding meets experimental ecotoxicology: advancing knowledge on the ecological effects of copper in freshwater ecosystems. Adv. Ecol. Res. 51, 79–104.

Graham, R.L., Hunsaker, C.T., O'Neill, R.V., Jackson, B.L., 1991. Ecological risk assessment at the regional scale. Ecol. Appl. 1, 196–206.

Green, S.B., 1991. How many subjects does it take to do a regression analysis. Multivar. Behav. Res. 26, 499–510.

Greenberg, A.E., Clesceri, L.S., Eaton, A.D., 1992. Standard methods for the examination water and wastewater., 18th ed American Public Health Association, American Water Works Association, Water Environment Federation.

Greenstein, D., Bay, S., Anderson, B., Chandler, G.T., Farrar, J.D., Keppler, C., Phillips, B., Ringwood, A., Young, D., 2008. Comparison of methods for evaluating acute and chronic toxicity in marine sediments. Environ. Toxicol. Chem. 27, 933–944.

Hajibabaei, M., Shokralla, S., Zhou, X., Singer, G.A.C., Baird, D.J., 2011. Environmental barcoding: a next-generation sequencing approach for biomonitoring applications using river benthos. PLoS One 6, 1.

Hampton, S.E., Strasser, C.A., Tewksbury, J.J., Gram, W.K., Budden, A.E., Batcheller, A.L., Duke, C.S., Porter, J.H., 2013. Big data and the future of ecology. Front. Ecol. Environ. 11, 156–162.

Hardy, C.M., Krull, E.S., Hartley, D.M., Oliver, R.L., 2010. Carbon source accounting for fish using combined DNA and stable isotope analyses in a regulated lowland river weir pool. Mol. Ecol. 19, 197–212.

Harris, G.P., Heathwaite, A.L., 2012. Why is achieving good ecological outcomes in rivers so difficult? Freshw. Biol. 57, 91–107.

Hedges, J.I., Stern, J.H., 1984. Carbon and nitrogen determinations of carbonate-containing solids. Limnol. Oceanogr. 29, 663–666.

Heino, J., 2013. The importance of metacommunity ecology for environmental assessment research in the freshwater realm. Biol. Rev. 88, 166–178.

Hilty, J., Merenlender, A., 2000. Faunal indicator taxa selection for monitoring ecosystem health. Biol. Conserv. 92, 185–197.

Huse, S.M., Welch, D.M., Morrison, H.G., Sogin, M.L., 2010. Ironing out the wrinkles in the rare biosphere through improved OTU clustering. Environ. Microbiol. 12, 1889–1898.

Johnston, E.L., Roberts, D.A., 2009. Contaminants reduce the richness and evenness of marine communities: a review and meta-analysis. Environ. Pollut. 157, 1745–1752.

Kelaher, B.P., Levinton, J.S., 2003. Variation in detrital enrichment causes spatio- temporal variation in soft-sediment assemblages. Mar. Ecol. Prog. Ser. 261, 85–97.

Kennish, M.J., 2002. Environmental threats and environmental future of estuaries. Environ. Conserv. 29, 78–107.

Kohli, G.S., Neilan, B.A., Brown, M.V., Hoppenrath, M., Murray, S.A., 2014. Cob gene pyrosequencing enables characterization of benthic dinoflagellate diversity and biogeography. Environ. Microbiol. 16, 467–485.

Long, E.R., 2006. Calculation and uses of mean sediment quality guideline quotients: a critical review. Environ. Sci. Technol. 40, 1726–1736.

Lücke, J.D., Johnson, R.K., 2009. Detection of ecological change in stream macroinvertebrate assemblages using single metric, multimetric or multivariate approaches. Ecol. Indic. 9, 659–669.

Magurran, A.E., Baillie, S.R., Buckland, S.T., Dick, J.M., Elston, D.A., Scott, E.M., Smith, R.I., Somerfield, P.J., Watt, A.D., 2010. Long-term datasets in biodiversity research and monitoring: assessing change in ecological communities through time. Trends Ecol. Evol. 25, 574–582.

McKinley, A.C., Miskiewicz, A., Taylor, M.D., Johnston, E.L., 2011. Strong links between metal contamination, habitat modification and estuarine larval fish distributions. Environ. Pollut. 159, 1499–1509.

Menezes, S., Baird, D.J., Soares, A.M.V.M., 2010. Beyond taxonomy: a review of macroinvertebrate trait-based community descriptors as tools for freshwater biomonitoring. J. Appl. Ecol. 47, 711–719.

Morgan, M.J., Chariton, A.A., Hartley, D.M., Court, L.N., Hardy, C.M., 2013. Improved inference of taxonomic richness from environmental DNA. PLoS One 8, e71974.

Olsgard, F., Somerfield, P.J., Carr, M.R., 1998. Relationships between taxonomic resolution, macrobenthic community patterns and disturbance. Mar. Ecol. Prog. Ser. 172, 25–36.

Pearson, T.H., Rosenberg, R., 1978. Macrobenthic succession in relation to organic enrichment and pollution of the marine environment. Oceanogr. Mar. Biol. Annu. Rev. 16, 229–311.

Porter, T.M., Gibson, J.F., Shokralla, S., Baird, D.J., Golding, G.B., Hajibabaei, M., 2014. Rapid and accurate taxonomic classification of insect (class insecta) cytochrome c oxidase subunit 1 (COI) DNA barcode sequences using a naïve Bayesian classifier. Mol. Ecol. Resour. 14, 929–942.

Quinn, G.P., Keough, M.J., 2002. Experimental Design and Data Analysis for Biologists. Cambridge University Press, Cambridge.

Reizopoulou, S., Simboura, N., Barbone, E., Aleffi, F., Basset, A., Nicolaidou, A., 2013. Biodiversity in transitional waters: steeper ecotone, lower diversity. Mar. Ecol. 35, 78–84.

Remane, A., 1934. Die brackwasserfauna. Verhandlungen der Deutschen Zoologischen Gesellschaft 36, 34– 74.

Rogers, H.M., 1940. Occurrence and retention of plankton within the estuary. J. Fish. Res. Board Can. 5a, 164–171.

Rygg, B., 1985. Distribution of species along pollution-induced diversity gradients in benthic communities in Norwegian fjords. Mar. Pollut. Bull. 16, 469–474.

Shokralla, S., Spall, J.L., Gibson, J.F., Hajibabaei, M., 2012. Next-generation sequencing technologies for environmental DNA research. Mol. Ecol. 21, 1794–1805.

Simpson, S.L., Spadaro, D.A., 2011. Performance and sensitivity of rapid sublethal sediment toxicity tests with the amphipod melita plumulosa and copepod nitocra spinipes. Environ. Toxicol. Chem. 30, 2326–2334.

Simpson, S.L., Batley, G.E., Chariton, A.A., Stauber, J.L., King, C.K., Chapman, J.C., Hyne, R.V., Gale, S.A., Roach, A.C., Maher, W.A., 2005. Handbook for sediment quality assessment. CSIRO, Sydney.

Sogin, M.L., Morrison, H.G., Huber, J.A., Welch, D.M., Huse, S.M., Neal, P.R., Arrieta, J.M., Herndl, G.J., 2006. Microbial diversity in the deep sea and the underexplored "rare biosphere" Proc. Natl. Acad. Sci. 103, 12115–12120.

Stark, J.S., Riddle, M.J., Snape, I., Scouller, R.C., 2003. Human impacts in antartic marine soft-sediment assemblages: correlations between multivariate biological patterns and environmental variables at Casey station. Estuar. Coast. Shelf Sci. 56, 717–734.

Stoeck, T., Bass, D., Nebel, M., Christen, R., Jones, M.D.M., Breiner, H.W., Richards, T.A., 2010. Multiple marker parallel tag environmental DNA sequencing reveals a highly complex eukaryotic community in marine anoxic water. Mol. Ecol. 19, 21–31.

Sun, M.Y., Dafforn, K.A., Brown, M.V., Johnston, E.L., 2012. Bacterial communities are sensitive indicators of contaminant stress. Mar. Pollut. Bull. 64, 1029–1038.

Sun, M.Y., Dafforn, K.A., Johnston, E.L., Brown, M.V., 2013. Core sediment bacteria drive community response to anthropogenic contamination over multiple environmental gradients. Environ. Microbiol. 15, 2517–2531.

Traunspurger, W., Drews, C., 1996. Toxicity analysis of freshwater and marine sediments with meio- and macrobenthic organisms: a review. Hydrobiologia 328, 215–261.

Underwood, A., 1991. Beyond BACI: experimental designs for detecting human environmental impacts on temporal variations in natural populations. Mar. Freshw. Res. 42, 569–587.

USEPA, 1996. Method 8260B Volatile Organic Compounds by Gas Chromatography/Mass Spectrometry (GC/MS). US Environmental Protection Agency, Washington.

Vitousek, P.M., Mooney, H.A., Lubchenco, J., Melillo, J.M., 1997. Human domination of Earth's ecosystems. Science 277, 494–499.

Wiens, J.A., 1989. Spatial scaling in ecology. Funct. Ecol. 3, 385–397.

Woodward, G., Baird, D.J., Dumbrell, A., Hajibabaei, M. (Eds.), 2014. Big Data in Ecology. Advances in Ecological Research, 51.

Zinger, L., Gobet, A., Pommier, T., 2012. Two decades of describing the unseen majority of aquatic microbial diversity. Mol. Ecol. 21, 1878–1896.

CHAPTER TWO

Big Data and Ecosystem Research Programmes

Dave Raffaelli*[,1], James M. Bullock[†], Steve Cinderby*, Isabelle Durance[‡], Bridget Emmett[§], Jim Harris[¶], Kevin Hicks*, Tom H. Oliver[†], Dave Paterson[||], Piran C.L. White*

*Environment, University of York, Heslington, York, United Kingdom
[†]NERC Centre for Ecology and Hydrology, Wallingford, Oxfordshire, United Kingdom
[‡]Cardiff School of Biosciences, Cardiff, United Kingdom
[§]NERC Centre for Ecology & Hydrology, Environment Centre Wales, Bangor, Gwynedd, United Kingdom
[¶]Environmental Science and Technology Department, School of Applied Sciences, University of Cranfield, Cranfield, United Kingdom
[||]Scottish Oceans Institute, East Sands, University of St. Andrews, St. Andrews, Scotland, United Kingdom
[1]Corresponding author: e-mail address: david.raffaelli@york.ac.uk

Contents

Abstract

The size and complexity of data sets generated within ecosystem-level programmes merits their capture, curation, storage and analysis, synthesis and visualisation using Big Data approaches. This review looks at previous attempts to organise and analyse such data through the International Biological Programme and draws on the mistakes made and the lessons learned for effective Big Data approaches to current Research Councils United Kingdom (RCUK) ecosystem-level programmes, using Biodiversity and Ecosystem Service Sustainability (BESS) and Environmental Virtual Observatory Pilot

Advances in Ecological Research, Volume 51
ISSN 0065-2504
http://dx.doi.org/10.1016/B978-0-08-099970-8.00004-X

(EVOp) as exemplars. The challenges raised by such data are identified, explored and suggestions are made for the two major issues of extending analyses across different spatio-temporal scales and for the effective integration of quantitative and qualitative data.

1. INTRODUCTION

Late at night, a police officer finds a drunk man crawling around on his hands and knees under a streetlight. The drunk tells the officer he's looking for his wallet. When the officer asks if he's sure this is where he dropped the wallet, the man replies that he thinks he more likely dropped it across the street. Then why are you looking over here?, the befuddled officer asks. Because the light's better here, explains the drunk. Moral: researchers tend to look for answers where the looking is good, rather than where the answers are likely to be hiding.

Freedmand (2010)

An internet search for the term "Big Data" reveals a myriad of definitions. "Large" and "complex" feature in most. Where data volume and data complexity and the interaction between them means that processing them, from their capture, curation, storage and analysis, synthesis and visualisation, becomes difficult and inefficient using traditional database management tools and applications, they become Big Data. Nowhere is this more true than for ecosystem-wide data sets collected over large geographical areas and which cover the physical, biogeochemical, biological, social and economic disciplines, with all the issues of data collection and storage standardisation of qualitative and quantitative approaches. In addition, these diverse data are usually collected at a range of spatial and temporal scales of grain (size of unit sampled, an area, volume or time interval), lag (distance between individual samples measured as a spatial or time interval) and extent (spatial area or time frame over which the samples are dispersed) (Raffaelli and White, 2013). Also, it is often the case that they do not share common spatial or temporal reference points that could serve as independent variables when searching for correlations and associations. The challenges facing large-scale, interdisciplinary research programmes concerned with the Ecosystem Approach to environmental sustainability (Convention on Biological Diversity, 2009), and which attempt to integrate data and processes from genes to landscapes and from across a range of disciplines, are non-trivial. Historically, the research community has dealt with large, complicated data sets by the simple expediency of reducing both size and complexity, through simplifying, pooling, averaging, summarising and generalising, thereby throwing away huge amounts of information and, potentially, missing opportunities to tease out signals from what has often been perceived as "noise". More often than not, that "noise" has been discarded because individual researchers saw no merit in its long-term curation, and if they did, they had no facilities or methodologies for doing so. Modern advances in information processing, data management and analytics means that these behaviours need not be tolerated in a world where the return on investment by tax payers for research needs to be transparent and accountable.

The analyses of Big Data can generate impressive relationships and patterns that provide tantalisingly attractive, often seductive, solutions to ecological and environmental issues (Aldous, 2011; Hampton et al., 2013; Michener and Jones, 2012), ranging from Beijing's air pollution,[1] to climate change and crop management.[2] But Big Data analysis has also drawn its criticisms, not least "The Street Light Effect" captured by the preliminary quotation for this chapter: we tend to explore the data we have, rather than the data we need, using the tools available, rather than the tools we need. Similarly, some of the Big Data analyses to date are unashamedly hypothesis-free, or even question-free, involving the mining of data sets to establish linkages and associations and then constructing a *post hoc* narrative without knowing anything about underlying cause-effect processes. In contrast, much of academic ecosystem-level research has been built on rigorous hypothesis formulation and testing. Nevertheless, an argument can be made for the power and potential of Big Data applications to reveal novel and unexpected insights from more inductive approaches that are quite theory-free. Extending the Street Light metaphor, one can occasionally look for something in the wrong place and indeed find something valuable.

Some of the challenges of Big Data in ecosystem ecology are immense and to help the reader grasp these, we pose the question "How might one design an ecosystem-level research programme that utilises Big Data approaches and frameworks?"

We ask the reader to suspend belief and imagine a global research programme that identified a range of diverse ecosystem types from around the world, each with its own core group of dedicated top-flight researchers collecting the same kinds of data and measuring the same fundamental processes

[1] http://qz.com/230689/how-ibm-is-using-big-data-to-fix-beijings-pollution-crisis/.
[2] https://www.facebook.com/microsoftresearch/posts/353849148007229?0hp=001b.

using standard and agreed methods minutely detailed in widely available handbooks written by that research community. Furthermore, analysing the underpinning ecological processes of key ecosystem services, such as food production, would be the goal of all the different groups working across the world in ecosystems ranging from the high arctic to the tropics and from marine to freshwater to terrestrial. The benefits of ecosystems for human well-being, captured as the health of the human populations within those ecosystems, would be evaluated. This diverse research programme would be co-ordinated centrally by individuals with a shared vision of the programme goals and the outcomes of individual projects would be synthesised into a greater whole for the long-term benefits of society. Most readers would probably agree that such a programme would indeed be visionary and, given the fact that another 3 billion people need to be fed by 2050 and that the Millennium Development Goals have yet to be realised with respect to many areas of health, such a programme would indeed be timely.

Fantasy? No. In fact, it is already been done, but few of today's generation of researchers are aware of the *International Biological Programme (IBP) 1964–1974,* even though they owe much of their craft and the fundamentals of their science to it. This ignorance is largely due to our inability at the time to cope with the Big Data challenges that the programme created. It may have been far ahead of its time in scope and vision, but the tools to make full and effective use of the data it generated were simply not available then. Informatics as a science was a glint in the eye of early pioneers and ideas of digital storage and processing were at a very early stage of development. The analysis of the individual systems looks simplistic from today's perspective, even though an agreed ideal goal was to represent them as stocks of natural capital and flows of materials within a systems analysis framework (Fig. 2.1). It is by exploring the origin and development of this endeavour that we can hopefully avoid recreating these issues.

2. THE IBP: A LESSON FROM HISTORY

The IBP (1964–1974) aimed to "ensure a world-wide study of . . . organic production on the land, in fresh waters and in the seas, and the potentialities and uses of new as well as of existing natural resources . . ." (Worthington, 1969). It recognised the need to feed a growing world after World War II, particularly in developing countries, and to base practical policies on a clear scientific understanding of the functioning of ecosystems, the relationships between stocks and flows of ecosystem components and their

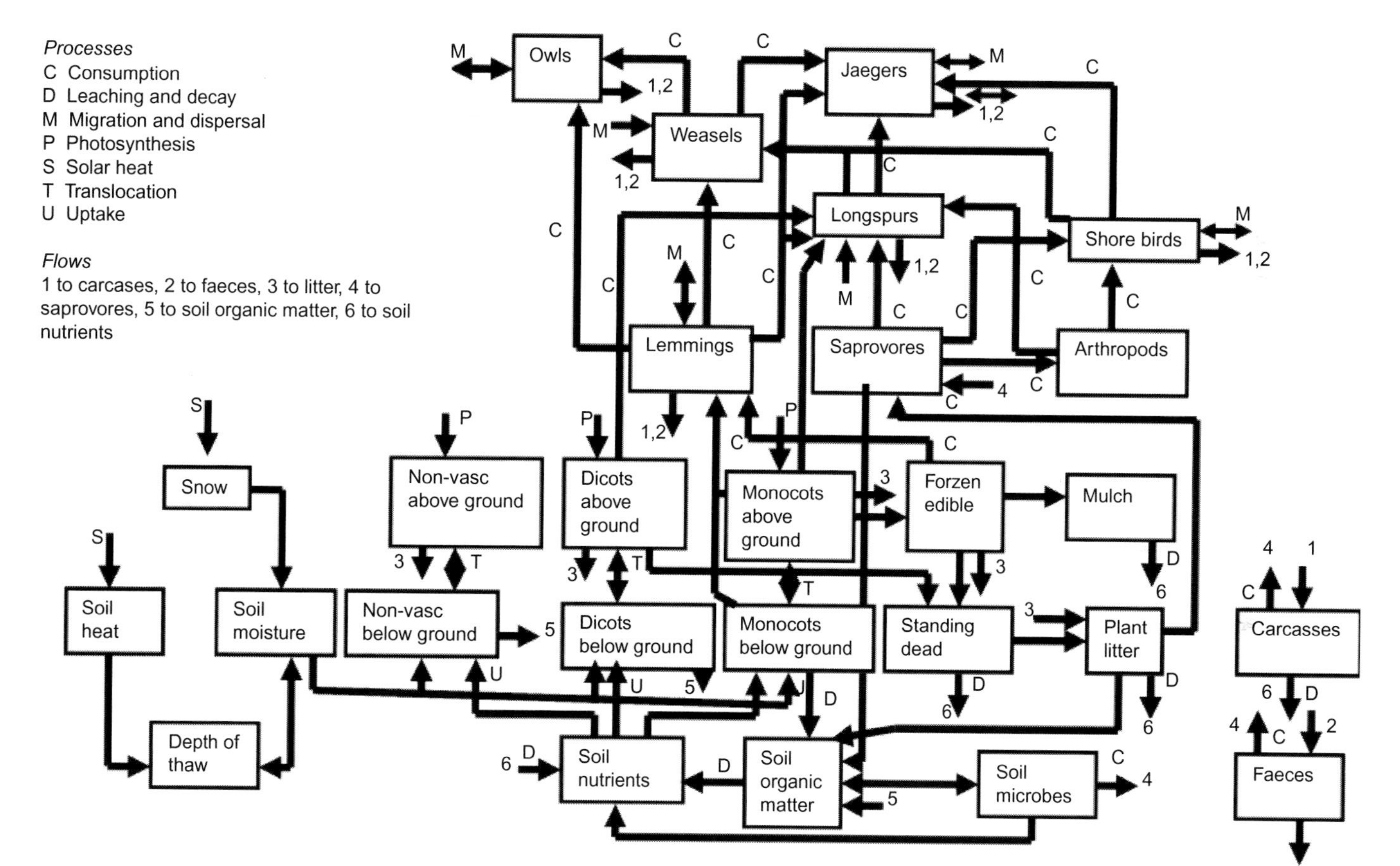

Figure 2.1 Box- and-flow diagram of a tundra ecosystem, Point Barrow, Alaska, typical of the representations used in IBP programmes to illustrate the relationships between key stocks of biomass. *Adapted from Worthington (1975).*

relation to human welfare (Worthington, 1969, 1975, 1983). This mission statement will be only too familiar to many of today's researchers and the vision of scientific endeavour and international co-operation presented in the contemporary reports, newsletters and briefings of the IBP are staggering by today's standards. More than 50 countries participated, and to give the reader an idea of the ambition and scope of the programme, Table 2.1 shows a list of just the shallow-water marine environments formally under the IBP banner. Acknowledging the need for a social dimension in ecosystem research was also anticipated by the IBP through its *Human Adaptability* section (Collins and Weiner, 1977), although, disappointingly, this soon became restricted to physiological aspects of health (Weiner and Lourie, 1969), despite pleas for a much broader social science research agenda (see Raffaelli and Frid, 2010). One of the most significant and lasting deliverables from the programme has been the "*Methods For*..." handbooks,[3] written in order to try to inject a degree of standardisation and comparability between studies, although researchers were never constrained to slavishly adopt these techniques, thus allowing their further development (Worthington, 1975). The programme had a finite life and although there are several synthesis volumes for specific regions (Worthington, 1975, Appendix 5), these represent the tip of the iceberg in what could have been achieved with today's know-how.

So, why was the full potential of the IBP not realised? Following the formal end of the programme in the mid-1970s, the IBP's two prime movers, the United States and the UK, reflected on what had been achieved and what lessons could be learned for future initiatives. At its peak, 1800 US scientists participated in the programme supported by $57 million in federal funds (Boffey, 1976), an astonishing amount by today's standards (equivalent today to $330 million by the Consumer Price Index or $874 million as relative share of GDP). However, there were concerns of "ecological sprawl" as projects became labelled as IBP which were probably marginal to the original goals and the US funders raised concerns about the lack of central governance of the science within the US and for the programme as a whole (Boffey, 1976). Finally, although much was learnt by the US ecosystem community as to how to work across the natural and physical sciences, one of the major science objectives, to develop systems analysis models of ecosystem to assess human impacts and predict the effects of natural change

[3] Worthington (1975) lists 24 of these published by Blackwells, Oxford, together with many regionally focused manuals in his Appendix 5.

Table 2.1 Selected examples of national projects focusing on marine shallow-water systems under the International Biological Programme umbrella. From Raffaelli (2000).

Programme	Location	Scientist responsible
Biology of the interstitial sand bottom fauna of the German North Sea Shore	Sylt. Heligoland, Germany	P. Ax
Metabolism of the benthic fauna investigated by radioactive tracer methods	North Sea, Germany	W. Ernst
Investigations on metabolism at the sea bottom of the German Bight	North Sea, Germany	S. Gerlach
Mytilus growth rates, dry matter production, variation in production of different races, transplantation experiments	Danish fjords, Denmark	B. Muus
Biology and productivity of the brackish water in the Baltic	Baltic coast, Finland	H. Luther
Dynamics of primary and secondary production of plankton and benthos	Mediterranean, France	
Food-chain, quantitative studies. Biomass and losses of food at different trophic levels	Dutch Wadden Sea, The Netherlands	J.J. Beukema
Benthic algae on intertidal flats	Dutch Wadden Sea, The Netherlands	G.C. Cadee
Food chains from benthic algae to young fish	Loch Ewe, Scotland, UK	J. H. Steele
Quantitative study of the food relations in an estuarine community	Ythan estuary, Scotland, UK	G.M. Dunnet
Primary productivity of mangrove swamps and lagoons	Cananeia, Sao Paulo, Brazil	C. Teixeira
Biological associations	Mundau lagoon, Brazil	L. Cavalcanti
Productivity in a small marine bay	St. Margarets Bay, Nova Scotia	L.M. Dickie
Organic production and the marine food chain	Cochin estuary, India	N.K. Panikkar and S.K. Qasim
Productivity of estuarine. Inshore and offshore marine environments	Panaji, Goa, India	N.K. Panikkar

Continued

Table 2.1 Selected examples of national projects focusing on marine shallow-water systems under the International Biological Programme umbrella. From Raffaelli (2000).—cont'd

Programme	Location	Scientist responsible
Comparative ecology of estuaries. Lagoons and mangrove swamps	India	R.V. Seshaiya
Comparative studies on the productivity of fresh, brackish and marine lagoons	South India	S. Krishnaswarni
Climate and hydrography of intertidal and shallow-water areas	Leigh, New Zealand	W.J. Ballantine
Productivity of *Spartina*	Auckland, New Zealand	V.J. Chapman
Biological reworking of marine sediments	USA	D.C. Rhoads
Ecology of eelgrass communities	Aleutian Islands, Alaskan Peninsula	C.P. McRoy
Ecological investigations of the flora and fauna of the littoral	White Sea, USSR	L.A. Zenkerich
Productivity of the high boreal littoral	Bering and Okhotsk Sea, USSR	O.G. Kusakin

Data extracted from IBP (1969).

"largely failed, primarily because the goal was unrealistic in view of the lack of valid theory and experience in dealing with such large and complex systems" (Boffey, 1976). This view is echoed in the British post-mortem chaired by Martin Holdgate (Worthington et al., 1976). The data demanded to construct system analysis models was underestimated and ability to use those data overestimated, while there was no effective repository for the huge amounts of data collected (Worthington et al., 1976).

One cannot help but be struck by the familiar ring of the issues identified. There was a lack of overall programme governance and an unwillingness of some disciplines to become engaged at the start and who therefore had little influence on the direction of the science as well as issues of working across the disciplines, especially between the natural and social sciences. In addition, there were few plans for data storage and management and for final synthesis. Finally, the programme was hindered by a tendency of groups to confine their vision to those ecosystems with which they are most familiar and comfortable (Raffaelli and Frid, 2010). In the following sections, we reflect on how far we have come with respect to these issues in ecosystem science research programmes, as illustrated by current effort by the UK's research community and with special reference to the *Biodiversity and Ecosystem Service Sustainability (BESS)* programme.[4]

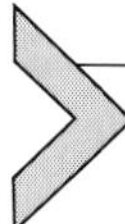

3. SIZE AND COMPLEXITY OF DATA SETS: ISSUES OF DISCIPLINE AND SCALE

Ecosystem ecology is now a much broader church than when the IBP dominated the field in the 1960s and 1970s. From the time the term "ecosystem" was coined by Roy Chapman and promoted by Arthur Tansley (1935), to as late as the 1990s (Raffaelli and Frid, 2010), mainstream ecologists continued to see ecosystems as entirely biophysical entities (e.g. Willis, 1997) with little attempt to include social dimensions. The Millennium Ecosystem Assessment (2005), with its emphasis on ecosystem services and human well-being, brought to a wide audience the sense that people are part of, not apart from, ecosystems, even though these ideas had been mainstream in some research communities many years earlier (Gómez-Baggethun et al., 2010). Ecosystem ecologists must now deal not only with physical, chemical and biological data and processes, but with information from the social and economic fields, if they are to understand the behaviour and dynamics of ecosystems, as argued below. This adds greatly to the complexity of data sets and has implications for their analysis and synthesis.

In research programmes like BESS, the data from individual component projects and work packages are not necessarily extremely large (although some extend to many terabytes), but they are complex. It is this combination of size and complexity which qualifies them as Big Data, in that they are difficult to capture, curate, store, analyse, synthesise and visualise. For the overall programme to be effective, these data also need to be integrated across component projects to address the larger BESS science questions, specifically:

- What are the relationships between biodiversity stocks and flows of ecosystem services?
- What are the important future drivers of change and how will these affect stocks and flows in the future?

[4] http://www.nerc-bess.net/.

- What tools and metrics can be developed as indicators to measure and track trends in stocks and flows?

When designing publically funded research programmes to answer specific questions, a tension arises between the need to constrain what researchers will actually do in order to avoid "ecological sprawl" (Boffey, 1976) and the need to allow the research community to be creative in how they approach the problem in open competition for funds. A compromise is to define the broad parameter space for the fundamental questions to be addressed and be flexible in how this is to be achieved by the successful component projects, picking the best science possible. While there are clear risks in relaxing control over exactly what data are collected and how they are collected, this is more than compensated for by the novel and creative research that emerges both at the initial competition stage, and also later over the life of the programme. Such a strategy demands trust and respect among all the research players and the directorate and for the BESS programme this has played out well. However, it does pose especial challenges for integrating data from the component projects into the larger programme data sets. Those data will not be exactly comparable so that Big Data analysis and synthesis within the programme requires a different approach, as discussed below.

The complexity of the data sets lies not only in their physical size, but in their diversity (quantitative, qualitative and all at a range of scales). Many biodiversity data sets are quantitative or semi-quantitative and usually recorded from point samples varying in grain from a few mm^2 to tens of m^2. If such measures are made over a large spatial extent, they are often presented as values within regular grid cells (e.g. breeding bird data, mammal sightings, insect records, weather data, and soil measures). In contrast, many socio-economic data are collected from irregular-shaped extents of varying sizes, such as local government administrative districts, individual farms or parishes. Both kinds of data can be transformed to surfaces which can then be linked to explore associations and correlations, but it should be remembered that these new surface data sets are merely models and summaries of the original underlying data, and field validation of these models is not common practice.

The need to work across different disciplines is more than just an article of faith, one of the miss-placed criticisms levelled against interdisciplinary research by some commentators. The added value of natural, physical, social and economic researchers working together to solve complex problems high on society's agenda has been ably demonstrated in the UK through the many outcomes of the *RELU Programme* (Lowe and Phillipson, 2006;

Lowe et al., 2009), the *UK National Ecosystem Assessment* (2011, 2014), *UK National Ecosystem Assessment* (2011) and recognised in international initiatives such as *Future Earth, Diversitas* and the *Inter-governmental Process on Biodiversity and Ecosystem Services (IPBES).* There remain substantial challenges to interdisciplinary working (Jones and Paramor, 2010) which make it all too easy to avoid doing so, not least of which is the incorporation of qualitative and quantitative data within the same analyses. Also misunderstandings and misinterpretations can occur when interpretation of data and processes derived by one discipline is made by those working in a totally different discipline, not only due to a lack of technical and linguistic understanding, but also due to different mental constructs and world-views. Interdisciplinary programmes that make no effort to broker respect and understanding between the different disciplines and to recognise these different world-views yield poor return on funded investment. Such bridges do not develop naturally and have to be built and nurtured through strong governance and leadership (White et al., 2009), something which is claimed to have been missing in the IBP projects (Boffey, 1976). The BESS programme has strong governance and leadership at both the directorate (overall programme) level and at the level of the main consortia of teams of researchers, and this has allowed better integration of different disciplines and of data management. Furthermore, empathy between the natural and social sciences has increased markedly since the IBP and interdisciplinarity is now almost a by-word in ecosystem science. For example, for the BESS consortia funding calls, about 15% of the total funding for the main consortia (ca. £10 million) was allocated to social science by the Natural Environment Research Council (NERC) that historically has been concerned purely with natural science, as well as through five interdisciplinary PhDs.

Interdisciplinary working creates real challenges with the curation, analysis and synthesis of the different kinds of data collected by the different disciplines. Much of the ecological data collected within BESS comes from an ecological community steeped in a rigorous, hypothesis-testing Popperian philosophy. The hypothetical-deductive approach is often viewed by social science colleagues as limited and constrained, being overly concerned with replication, control and experimental falsification, while the more empirical and freer explorations of social scientists can be anathema to experimental ecologists. Where these different world-views are held within BESS, they are seen as healthy tensions, not barriers to progress, and their resolution can lead to new ways opportunities and ways of thinking and understanding by the different disciplines.

Historically in UK science, data from different disciplines have been curated in different repositories. For example, social science and interdisciplinary data are held in the UK Data Archive,[5] while ecological and environmental data are held in several different archives depending on whether they are marine (British Oceanographic Data Centre, BODC[6]), terrestrial and freshwater (Environmental Information Data Centre, EIDC[7]) and there is an Earth Observation Data Centre.[8] Having separate and distinct data centres has in the past meant that the full potential of Big Data has not been realised but most of these data centres now have portals to the others. It has been a condition of BESS funding that all component projects submit their data to the EIDC so that there is a common repository for all the programme data and the projects are supported by dedicated EIDC staff.

Developments are in place through the UK's recent environmental data investments targeted on Big Data, including increasing the capacity of the JASMIN (Java Virtual Machine) National Data Computing Facility and the Climate, Environment and Monitoring from Space (CEMS) Earth Observation facility, developing the cloud-based technology demonstrated by the Environmental Virtual Observatory (EVO) programme (see below) and digital conversion of historical data.

4. THE BESS PROGRAMME

BESS is a 6-year NERC programme (2011–2017) designed to deliver the fundamental science on the linkages between natural capital stocks and flows of ecosystem services, framed around the three core questions listed above. To answer those questions, BESS uses a diverse range of delivery mechanisms: 4 consortia, 7 smaller research grants and fellowships, a vibrant PhD (15) and post-doc (24) community, together with workshops and working groups that provide the glue to hold the network of ca. 170 researchers and ca. 50 major stakeholders together (Figs. 2.2 and 2.3). The consortia provide the foci for the grants, studentships and workshops and each works in a different landscape type (lowland agriculture—Wessex BESS (Wessex Biodiversity and Ecosystem Service Sustainability); upland rivers—DURESS (Diversity of Upland Rivers for Ecosystem Service Sustainability); urban areas—F3UES (Fragments, Functions, Flows and Urban

[5] http://www.data-archive.ac.uk/.

[6] http://www.bodc.ac.uk/.

[7] http://www.ceh.ac.uk/data/.

[8] http://neodc.nerc.ac.uk/.

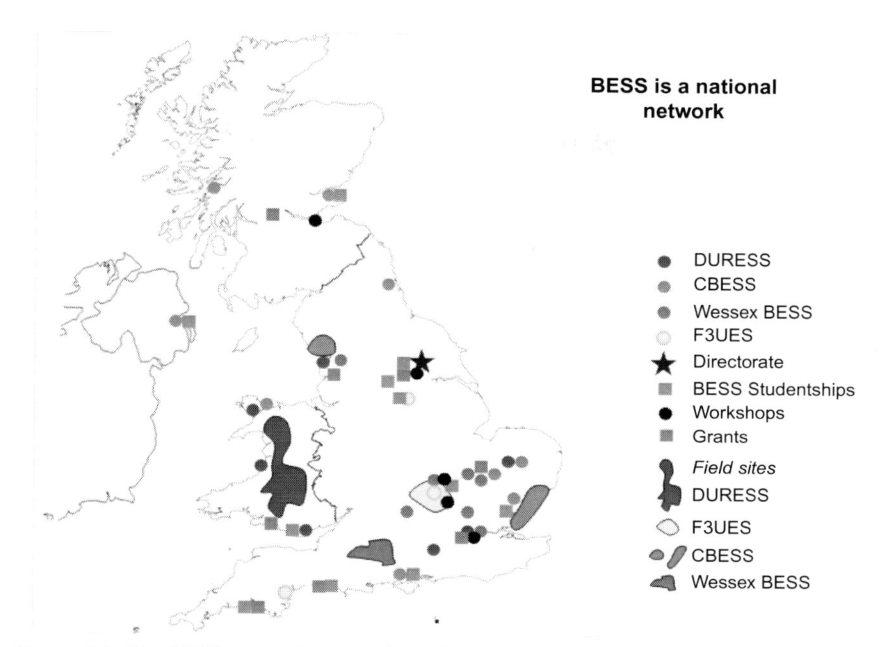

Figure 2.2 The BESS research network and main consortia field sites.

Ecosystem Services); coastal wetlands—CBESS (Coastal Biodiversity and Ecosystem Service Sustainability)). What follows are personal reflections of the principle investigators that lead those consortia on the issues that Big Data present for them.

4.1. DURESS

DURESS focuses explicitly on upland UK rivers and their landscapes (Figs. 2.2 and 2.4). The UK has over £250 billion invested in water infrastructure and the water industry contributes around £10 billion annually to the UK economy. This makes the UK's 389,000 km of streams and rivers one of its most important natural assets. However, the value of that asset is often overlooked, in particular, the processes carried out by the multitude of river organisms that together help to maintain and regulate water quality, for example by processing organic carbon and nutrients. These same organisms are part of an intricate food web that fuels everything from Atlantic salmon to birds like dippers and kingfishers, yet very little is known of how different parts of this web fit and function together. We also lack quantitative understanding of how these river processes contribute to delivering the key ecosystem services on which society relies: not only clean water, but

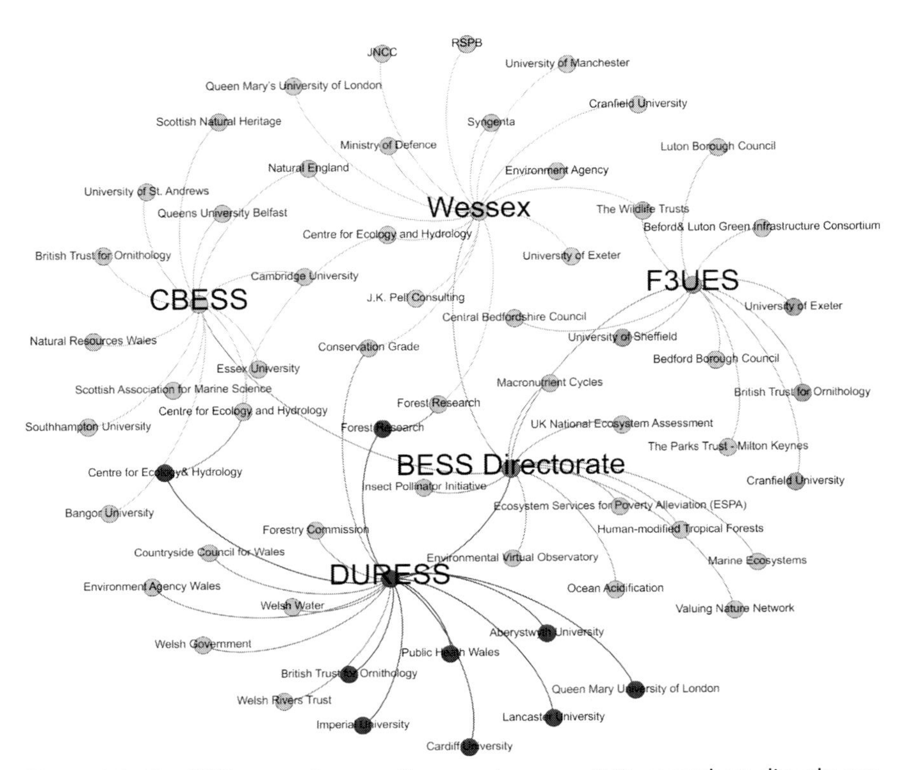

Figure 2.3 The BESS network currently comprises over 170 researchers directly connected with the research projects, plus early career researchers and those participating in workshops and working groups. The network diagram below captures the linkages between institutions, with the major stakeholders and with other RCUK programmes.

also the fisheries enjoyed by Britain's 4 million anglers, or river birds that are such an important part of our culture and our enjoyment of freshwater landscapes. DURESS is thus filling critical knowledge gaps in which there is major stakeholder interest. Addressing these research needs means investigating river microbes, invertebrates, fish and birds at levels of organisation from genes to food webs to test the overarching hypothesis that: "Biodiversity is central to the sustainable delivery of upland river ecosystem services under changing land-use and climate". Efforts are focused on river ecosystem services that are explicitly biodiversity-mediated, including the regulation of water quality, the production of fish for fisheries and recreational fishing, and river birds as culturally valued biodiversity. Each is at risk from climate/land-use change, and all require an integrated physical, biogeochemical, ecological and socio-economic science perspective that none

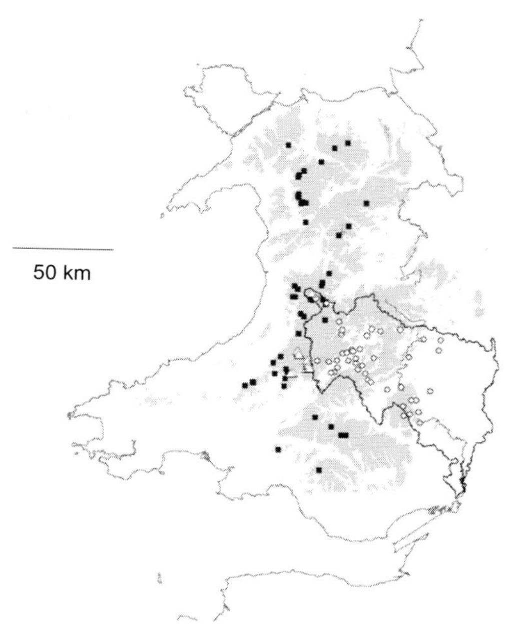

Figure 2.4 Map of Wales, UK, showing available data for the DURESS consortium. Black squares indicate the Welsh Acid Waters sampling points, open triangles the Llyn Brianne sampling points and open circles the River Wye sites. Areas above 200 m in altitude (uplands) are shaded in grey. Contains Ordnance Survey Data, Crown Copyright and Databaseright 2011.

of the project partners could deliver alone but which in turn generates large and complex data sets. In addition, there is a need to derive evidence over all scales at which ecological systems/services are organised, recognising that ecosystem functions and services are scale-dependent. Finally, we need to embed temporal dimensions relevant to ecosystems and to their drivers of change. In response to these complex needs, DURESS has created a hierarchical nested framework for the rationalisation and organisation of Big Data that takes advantage of the natural nestedness of river catchments. Within this framework, three Big Data sets were built covering a range of spatial and temporal scales (see Box 2.1).

To maximise our analytical power, a range of strategies for acquiring the data are adopted: (1) collation and harmonisation of existing historical data to create spatially, temporally, and thematically extensive data sets; (2) Update of historical data sets with contemporary data using identical protocols, and filling in gaps using an array of modelling techniques; (3) where historical

BOX 2.1 Extent of DURESS Big Data

Three "Big Data" sets were assembled covering a range of spatial and temporal scales.

a. A large-scale, coarse resolution historical data set covering most of the British uplands: in more than 1000 sites, 23 years of annual information on water quality, invertebrate families, fish, birds, land-use change were extracted from public monitoring schemes (e.g. WFD monitoring data, Bird Atlas, EU Land mapping). Challenges here lie in finding sufficiently long-term series of appropriately high quality, and in selecting, storing and then analysing extremely large data sets. For example, to collate the fish component of this data set, selection was made from a database of 3.3 million individual fish size/lengths.
b. A stratified intermediate resolution in representative catchments rich in existing historical data. Complementary contemporary data covering all the river organisms from microbes to invertebrates, fish and birds, across all levels of organisation from gene to food web, were collected from 50 catchments across upland Wales. This data set includes a range of measurements from classical measures of biodiversity for which historical values were available, to more novel measures based on energy flows, efficiency of resource partitioning, trophic linkages, competition, genetic traits. Overall, more than 18,000 diatoms, 245 biofilms, 34,786 invertebrates, 15,951 fish and 240 birds were sampled, weighed, identified and assessed.
c. A smaller scale/fine resolution in upland sub-catchments typifying contrasting upland use. Experimental data, the "golden standard" of natural scientists, are collected at reach scale and at fine temporal resolution (15 min to a few days). Here, the costs lie in the need for state-of-the-art measuring tools and techniques (e.g. field spectrolysers, stable isotopic analyses, genetic analyses) that can provide accurate high resolution data.

data are insufficient to inform the research questions, more relevant novel data are collected, but from the same historical locations; and (4) whenever possible novel data collection methods are standardised with those of other on-going programmes (e.g. the NERC *Macronutrient Cycles* Programme).

The main challenges encountered in exploiting Big Data for this project lie in: (i) the capacity to characterise the level of uncertainty or inaccuracy that stems from the analysis of Big Data; (ii) the dependency on historical data to be able to predict how future global changes may drive biodiversity and ecosystem service responses; (iii) the capacity to assess upstream—downstream dependencies specific to river networks.

It is often posited that Big Data may become as important to society as the Internet, the underlying assumption being that more data will lead to more accurate analyses and more confident decision making, leading to increased well-being and prosperity. While Big Data analysis certainly provides a unique opportunity for an integrated approach to real world complex problems, its value is limited by the accuracy of the data analysed and the capacity to measure the associated uncertainty in the outputs. There are high costs involved in assessing and recording data accuracy, namely in ensuring sufficient quality control and metadata filing. In DURESS, both the natural scientists and the social scientists are working hard to address these constraints, by working through a shared database management system, and by investigating new tools to assess uncertainty. For example, novel time-series tools are being tested to monitor uncertainty in the link between changes in catchment hydrology and river biodiversity, and new ecosystem service valuation methods are being developed that offer the opportunity to account for uncertainty and gaps in ecological knowledge.

One of the key aims of DURESS is to identify potential resilience and thresholds in service delivery in order to develop predictive tools for decision making. The project is therefore dependent on quality historical data. While originally collected in order to address other challenges (e.g. in response to European Water legislation and research on acidification), historical data collected over the last 20–30 years spanning spatial scales ranging from sub-catchments to whole regions are widely available in the UK, as is the case in many European countries. These extensive, but often compartmentalised data sets offer a unique opportunity to investigate ecological problems such as the role of river biodiversity in sustaining ecosystem services in a changing world. However, the approach is not without its challenges. The major limitation to historical data is that they are often limited in time, rarely spanning more than two decades, and therefore significantly limited in predictive capacity. In addition, historical data are rarely adapted to respond to contemporary ecosystem questions. For example, while family-level invertebrate Water Framework Directive (WFD) monitoring records can inform on the abundance of food resources for fish, they are of insufficient taxonomic resolution to explore other important aspects of biodiversity that are likely to be significantly linked to ecosystem services, such as functional traits or genetic make-up. DURESS researchers are exploring creative ways to work around these challenges, for example, by building models from new data to compensate for limitations in historical data.

One of the recurring challenges with analysing river data lies in the network structure of river systems which creates strong upstream—downstream dependencies. This is compounded for studies on ecosystem services because the beneficiaries of services are often located downstream of river catchments. Trout fishing, one of the ecosystem services assessed in DURESS, provides a good illustration of this problem. Anglers catch large fish that are predominantly downstream of upland catchments where the nurseries are located. The quality of this recreational ecosystem service lies significantly in the abundance and in the dynamism of fish caught downstream. However, this service is strongly determined by the processes that occur upstream. Since heavier than average juvenile fish in the upland nurseries are those that are most likely to survive and contribute to downstream migration, the angling ecosystem service can only be assessed by measuring specific characteristics of both downstream adult populations and upstream juvenile populations. These temporal or spatial dependencies in the Big Data add additional complexity to the analytical work, and offer significant statistical challenges.

4.2. CBESS

The CBESS consortium is the transitional terrestrial-marine habitat component of the larger BESS project. Coastal areas provide a diversity of ecosystem functions and services and CBESS is exploring these in the context of the three core BESS questions (see above) at a range of field sites around the UK (Fig. 2.2). What each member of the CBESS consortium does is relatively straightforward and at first sight the programme may seem surprisingly simple. The objective is to measure biodiversity, related functional variables, examine socio-economic drivers, responses and management tools for a number of coastal sites (salt marshes and mudflats) across the UK. These data will then be used to investigate the relationship between biodiversity stocks and the flow of ecosystem services. However, the devil is in the detail and to understand the CBESS challenges with respect to Big Data that detail needs to be described. Firstly, the CBESS team agreed to match measurements in space and time, all groups working where possible within pre-selected metre square quadrats, allocated by computer programme, randomly (almost) within a polygon describing the "boundaries" of each area of interest. Trial studies quickly established that there would need to be a clear hierarchy and control of measurements, largely based on their relative effect on the physical integrity of the sample site. Measurements were ranked according to the level of disturbance to the habitat ranging from non-invasive sampling (spectral reflectance, photography) to those measurements causing minor impact (surface stability and small cores) and finally to major disruptive sampling (large cores for macrofaunal sampling and flume studies). A hierarchical sequence of 18 measurements has been established and this field protocol is applied to each quadrat (Fig. 2.5). Quadrat-specific data are supported by site-wide measurements of atmospheric conditions, eddy covariance studies of carbon flux and information on wave dynamics and water depth. The numbers, logistic limitations and effort mount quickly. Most measurements require replication, and often five measurements are taken from each quadrat. Twenty-two quadrats are the best compromise for maximising sample size and minimising effort and logistics given that in each regional

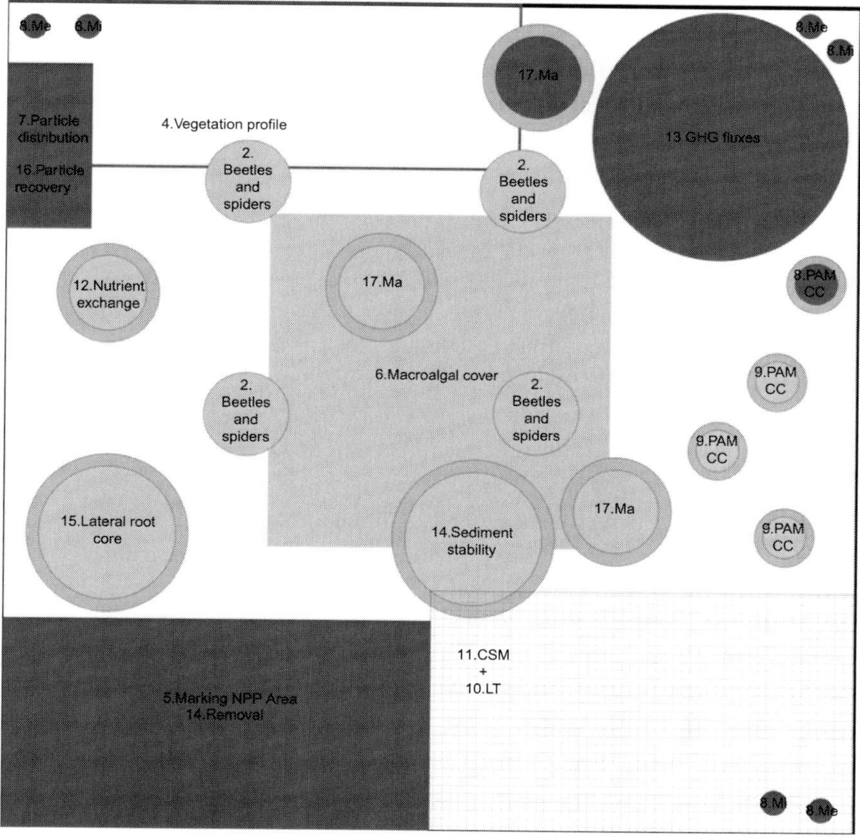

Figure 2.5 Standard sampling quadrat for different kinds of data by CBESS.

area (Morecambe Bay and Essex marshes) three mudflats and three saltmarsh habitats are sampled.

This sampling approach gives a basic background of 2640 measurements for a single variable multiplied by 18 for the different variables measured, yielding a theoretical >47,000 data points. However, sheer logistic and technological limitations means that some of the 18 measures have had to be truncated. In addition, many data are not single data points, but rather a species list or reflectance spectra, digital images and process measures (such as primary productivity and nutrient flux). Indeed, the molecular data (bacterial, archeal and meiofaunal diversity) on their own will far exceed that value of 47,000 while background data such as eddy covariance and pressure sensor data (wave conditions) have been collected continuously for more than a year. The data are now so big that they are better expressed in digital storage units rather than data points or sets.

However, the volume of data is not really the issue but rather, as in all Big Date challenges, how to relate data sets of such different characteristics and search for relationships among the groups of data to answer our basic questions. For example, molecular analysis provides an assessment of the numbers of operational taxonomic units within a bacterial sample. These numbers swamp the information gained on the diversity of the macrofaunal organisms recovered from traditional cores samples. In order to approach analysis of this combination of data and diversity metrics CBESS has formed a Data Analysis Working Group specifically to address these issues and explore different analytical opportunities. However, the complexity of the task is also increased by that fact that we also wish to consider scale and context (season and location) adding further complexity. The 22 quadrats placed within the selected site are not randomly assigned but contain a hierarchical relationship between samples based on scales of separation or lag (1, 10 and 100 m). The locations themselves then serve as comparators at super-km scales while the sites across the country represent landscape and super-regional scales. The design of this programme and the subsequent statistical approach has been (and still is) hotly debated within the project.

In addition to the biological and physical data, a whole new area of analysis is required in terms of the socio-economic component of the CBESS work. The socio-economic team have conducted workshops and interviews and data have been collected using state-of-the-art methods such as interactive digital imagery with real-time data input. This work is critical to inform our interpretation of how the public and stakeholders view biodiversity and its affects and is one of the goals of CBESS.

4.3. Wessex BESS

The Wessex BESS Study area is a lowland agricultural landscape centred in Wiltshire, UK (Figs. 2.2 and 2.6), in which we are studying the ecosystem services of water quality, food provision (pollination, pest control, forage production), climate regulation and cultural services.[9] The key focus of the project is to understand the mechanistic relationships between biodiversity and ecosystem services under differing environmental conditions. This mechanistic understanding will be subsequently used in combination with a variety of data sources to map ecosystem services under future land-use scenarios.

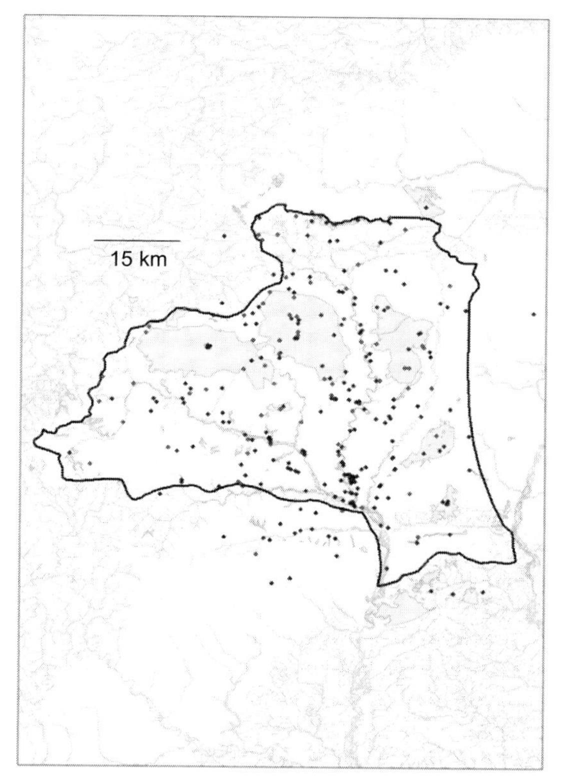

Figure 2.6 Wessex BESS example of spatial data reported by stakeholders as important cultural locations. Designated statutory protected areas (SSSIs) are shown in orange (light grey in print version). *Data credit: H. King.*

[9] http://www.brc.ac.uk/wessexbess/.

The data used in the project are both novel (e.g. gas flux measurements) and historical (e.g. water chemistry monitoring data, species occurrence records). Data quantities vary from small data sets (e.g. species lists of plants on experimental plots <1 megabyte) to medium (earth observation data from aerial flights >4 terabytes). The formatting of data varies from simple spreadsheets to more complex Oracle databases and processing of LiDAR (Light Detection And Ranging) earth observation data. In addition to environmental data, qualitative data from social science interviews are collected as part of this interdisciplinary project. Top copies of data are stored with respective host institutions (Wessex BESS is a collaboration between five institutions), but all data will be sent for archiving after the project to the NERC Environmental Information Data Centre, including associated metadata. As is the case for all the BESS consortia programmes and project, these data will be openly available after a 2-year embargo in order to allow data owners to publish their findings.

Oft-cited challenges of Big Data include the technical capacity to store and interrogate very large data sets (Howe et al., 2008). This is especially relevant to research fields such as molecular biology and earth observation where high-throughput analytical facilities and increasingly high spatial and temporal resolution measurements result in high volumes of data. However, the data issues for biodiversity-ecosystem service research in the Wessex BESS project are less related to size and storage (amounts of data actually being small compared to other Big Data research fields), but relate more to their complexity, that is, the integration of very disparate data sets across multiple disciplines. In particular, for the purpose of modelling ecosystem services, data sets are drawn from public health, earth observation, environmental monitoring and land-use planning disciplines among others. For a single individual, or modelling team, simply knowing about the existence of data and accessing data sets in a timely way can be a challenge. These issues are likely to be faced by many scientists engaged in modelling of ecosystem services and there is therefore the potential for a large amount of redundant effort spent by individual groups seeking similar data sets.

To overcome these problems, the creation of "data gateways" to search and locate useful data sets is of great value. This effectively centralises the tasks of locating relevant data, and reduces redundancy in multiple researchers carrying out the same task. A number of data gateways have developed in recent years (Section 3) and there is an extensive effort to upload useful data sets from previous projects to make these available with associated metadata which is fully searchable. These are excellent endeavours, but, as with all search engines, the user needs have some idea of what they are searching for. In some cases, researchers may not realise that different types of potentially useful data exist.

In order to model ecosystem services, the key input data needs are either data layers to parameterise process based models (e.g. river networks and topography for InVEST nutrient retention models; Kareiva et al., 2011; Nelson et al., 2009) or layers which themselves are direct proxies of ecosystem services (through a benefits transfer approach; Eigenbrod et al., 2010). However, as with any predictive modelling, there is also a critical need to validate results. A range of data sets already exist that may help achieve this (e.g. monitoring locations for water chemistry or counts of pollinating insects). In many cases, researchers may be unaware of the existence of key input and validation data sets. Therefore, in addition to searchable data gateways, a significant advance in the use of data for ecosystem service modelling could be gained through the development of metadata tables describing different possible input and validation data sets.

4.4. F3UES—Urban BESS

The F3UES consortium dealing with fragments, functions and flows is focussed on the delivery of ecosystems services in urban areas,[10] where most people live and experience ecosystem service delivery, a rapidly increasing trend globally. In order to fully understand how services are delivered in these areas in order to secure sustainable improvement in human well-being, it is necessary to integrate data ranging from the purely biophysical (e.g. soil carbon), through human impact factors (e.g. noise) through to assessment of subjective parameters (e.g. aesthetic appreciation). These data are organised in a hierarchy of spatial scales. The challenges working across disciplines to produce common approaches to data integration are legion, with issues of timing, location and, of course, the vagaries of the weather. F3UES is aimed at discovering how differing forms of urban green space deliver ecosystem services, what their Natural Capital status is and how this controls and sustains flows of energy, materials and biodiversity elements (e.g. birds, people) between the various fragments.

The project is addressing the major BESS questions (see above) using three urban areas each of which brings a distinctive set of urban form. These areas are: Milton Keynes, a planned New Town predicated on water management issues in addition to the usual infrastructure considerations; Luton,

[10] http://bess-urban.group.shef.ac.uk/.

a mixture of Victorian Terrace and industrial development; Bedford, the County Town of Bedfordshire, with an urban landscape that can be traced back to medieval times. It is important to stress that these three towns are not replicates, rather they merely provide a full range of green space forms of differing topology from small back gardens to large public open spaces to inform our relationships about function and flows between these fragments.

The collection of data involves a number of modes of acquisition. Remote sensing using the NERC-Airborne Survey and Research Facility (ASRF) for multi-spectral, LiDAR and visual data has generated over 1400 terabytes of data alone, and these data are linked to ground survey of "focal point" fragments, classified on the basis of GIS mapping to secure the whole range of fragment type. The integration of the ground survey with the remote data is a complicated task, and is on-going. The ground survey of fragments entails collecting invertebrates, plant data and materials, soils, hydrological measurement, bird data and thermal and noise data, allied to questionnaires on aesthetic and other human factors. This has required close collaboration between a wide range of disciplines. The survey of birds, led by the British Trust for Ornithology (BTO), has required the input of large numbers of expert volunteers, so that ensuring data robustness and fidelity requires close attention to detail.

Data are managed by building up common data sharing platforms between members of the consortium, with all data referenced (where possible) to location. This will be integrated and modelled using a diversity of approaches, including Bayesian statistical methods, and outputs will range from graphical representations of X–Y relationships to information rich maps, retaining the fine grain of the original data. This is essential if we are to map layers and secure integration for novel outputs, enabling us to ask questions such as "how are bird numbers and flows affected by soil carbon distribution?" or "how does manipulation of biomass resources affect movement of people and human factors?" This is critical if we are to derive useful information for populating tools that will aid decision making for planners operating in urban contexts, providing guidance for green space establishment and management for both new urban extensions, new towns, and retrofitting and managing long established urban areas. There is a plethora of work emerging showing that access and exposure to green space have positive mental and physical health outcomes (e.g. Natural England, 2012). However, for the planning community to act upon this they need robust models derived from sound data, and a mechanistic understanding of the interrelationships modelled.

F3UES faces a number of challenges outside those which one would normally encounter working in a multi- and interdisciplinary project. Principal among these is the sheer number of stakeholders involved and the potential problem associated with gaining access to the most appropriate fragments, although Local Councils, volunteer groups, businesses and hundreds of private individuals have generally welcomed us in to sample their gardens and green spaces. Sometimes access proves impossible and then a substitute area needs to be identified. Taking all three towns together, F3UES has in effect over half a million stakeholders which presents quite a challenge, especially when trying to achieve unbiased responses to questionnaires. Keeping these groups informed and on-board is a particularly challenging task, especially when some are not comfortable with disruption of the status quo when performing biodiversity manipulation experiments. This would not have been possible without the close co-operation of the respective Local Authorities and non-governmental organisations involved so that we have generally had a very positive reception from the large majority of local residents. The rewards from this very public project are potentially extremely high, and the potential for producing tools to put in the hands of individuals, in the form of telephone-based apps, is a particularly exciting prospect.

4.5. Summary of consortia Big Data challenges

It is clear that the BESS consortia, and the multitude of sub-projects which feed into each, are drawing on both historical and novel data that are large and complex (qualitative and quantitative and at a range of temporal and spatial scales). Each of the consortia has been given the freedom to choose tools and frameworks to address these issues independently in order to explore the potential of competing approaches and in acknowledgement of the different kinds of data collected and analysed. Nevertheless, consistent features emerge. All the consortia are using a rigorous hierarchical framework for sampling and collection, based on perceived relevant spatial scales, which should facilitate drawing these diverse data sets together at the Directorate level. All consortia have a significant socio-economic dimension, vital for understanding the behaviour and dynamics of their own system, but it is not clear yet how these data are going to be melded across the consortia for inclusion in Big Data analyses and visualisations. This will be an exciting challenge for BESS, as it is for all large-scale programmes, and we reflect on possible approaches towards the end of this chapter. In this respect, we now describe two important advances that have great potential for meeting those challenges: data and modelling platforms and novel ways to visualise complex data and results.

5. PLATFORMS FOR BIG DATA: THE ENVIRONMENTAL VIRTUAL OBSERVATORY

While never part of the BESS programme, the Environmental Virtual Observatory Pilot (EVOp) stand-alone project was, along with BESS, part of NERC's portfolio of activity designed to tackle some of the challenges associated with large-scale, holistic data collection and modelling of the kind demanded by an Ecosystem Approach to environmental management. EVOp ran in parallel with BESS with many researchers and institutions common to both programmes. Recently BESS has co-funded development of many of the activities and products that emerged through EVOp. From the following account of the EVOp, it is clear that the BESS projects will need to take advantage of the outcomes.

EVOp explored the implications of the unexpected outcome of the explosion of environmentally relevant data and associated information (Fig. 2.7). With this growth comes a growing disconnect between and within the supply of scientific knowledge and the demand for that knowledge from the private and government sectors. NERC commissioned the EVOp to explore the question: "Is there a way of providing the 'wiring' to help people access what resources they need, be they a scientist, policy maker, industrial body, regulator or public?" The hypothesis was that novel cloud-computing technologies could be exploited to increase accessibility in a data-intensive world in order to filter and integrate this information to manageable levels as well as provide visualisation and presentation services (Fig. 2.8) to make it easier to gain creative insights and build collaborations: the so-called the 4th paradigm of Big Data (Gray, 2007; Hey et al., 2009). The ultimate aim was to make NERC science more efficient, effective and transparent. The transparency issue had been highlighted in a recent Royal Society report as requiring urgent attention to increase public confidence in the process of translation of scientific evidence through to policy making (The Royal Society, 2012).

A second hypothesis was to test if there is value going beyond data to include models which are effectively a synthesis of our current understanding and one of the main tools NERC scientists use to integrate complex data, upscale and make projections under future scenarios. The resonances with

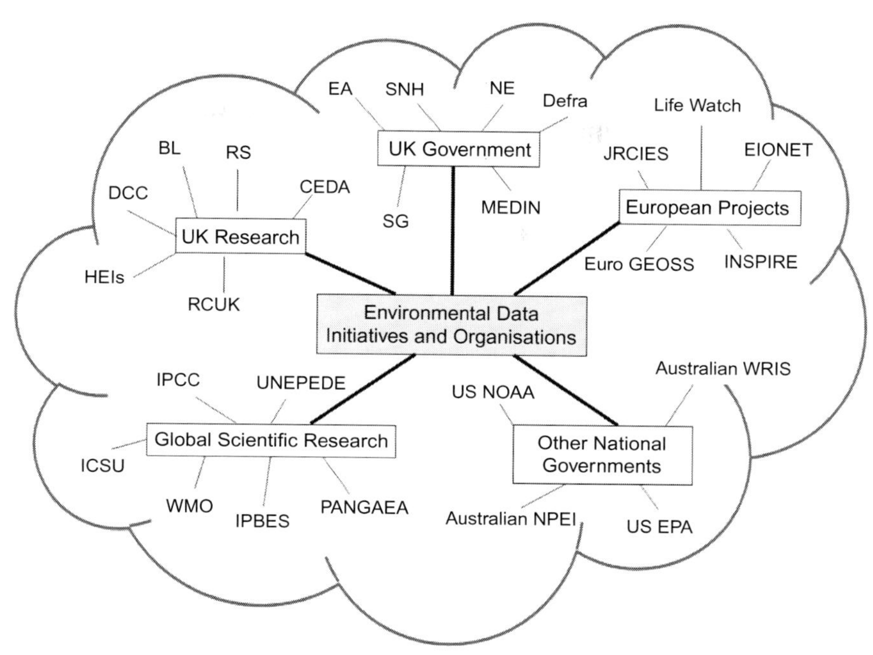

Figure 2.7 Examples of institutions that provide the environmental data landscape for virtual observatory platforms such as the EVOp. Key to acronyms: *UK Government*—EA Environment Agency; SNH Scottish Natural Heritage; NE Natural England; Defra Department of Food and Rural Affairs; SG Scottish Government; MEDIN Marine Environmental Data and Information Network. *European Projects*—INSPIRE Spatial Information in the European Community; JRC IES Joint Research Centre Institute of Environment and Sustainability; EIONET European Environment Information and Observation Network; EuroGEOSS European Group of Earth Observation System of Systems. *UK Research*—RCUK Research Councils United Kingdom; HEIs Higher Education Institutions; DCC Department for Climate Change; RS Royal Society of London; BL British Library; CEDA Centre for Environmental Data Archival. *Global Scientific Research*—IPCC Intergovernmental Process on Biodiversity and Ecosystem Services; UNEP EDE United Nations Environment Programme Environmental Data Explorer; ICSU International Council for Science; IPCC Intergovernmental Panel on Climate Change; WMO World Meteorological Organisation. *Other National Governments*—US NOAA United States National Oceanic and Atmospheric Administration; US EPA United States Environmental Protection Agency; Australian NPEI Australian National Plan for Environmental Information; Australian WRIS Australian Water Resources Information System. *Adapted from NERC (2014).*

the needs of the BESS consortia and other ecosystem programmes are obvious. Within the terrestrial and freshwater communities many models address environmental questions concerning soil and water quality, flood and drought risk, as well as ecosystem structure and function. These models simulate complex physical, chemical and biological process interactions. There

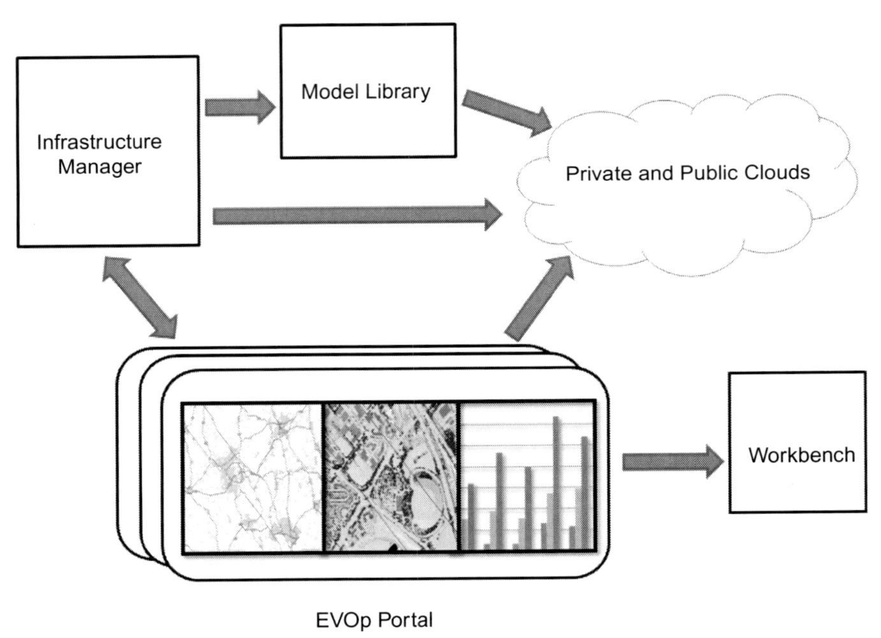

Figure 2.8 Schematic representation of the overall architecture of the EVOp. *Adapted from NERC (2014).*

is a challenge however in gaining access to these models, linking them together to deliver more holistic outputs and objectively testing to the level demanded by end-users who need to make policy or management decisions based on their outputs. This leads to the second question: "How can we create a culture of more open and rigorous testing and evaluation of the current models necessary to improve process understanding, process representation in models and thus model forecast accuracy?" Two main challenges currently limit this on-going model development: the spatial/temporal limitations of our observational capacity and the lack of integration of data, models and visualisation tools across the air–land–water domains.

The Agile method of end user-orientated system design and development was used to ensure end-user needs were built into the cloud infrastructure and data services from the very beginning. Storyboards were used to elaborate user requirements to test the efficacy of the cloud computing-enabled infrastructure to evaluate environmental behaviour and its likely response to changing climate and land use at a range of spatial scales. The team combined a "broad and shallow" approach for certain aspects such as intellectual property rights, licencing and regulatory issues with a "narrow and deep" method for developing scenarios covering river discharge predictions, uncertainty about climate change projections, land management interventions and diffuse water pollution at three spatial scales (local, national and global). These approaches ensured that the exemplars were tightly focused on real end-user questions an approach that was welcomed by organisations dealing with these complex problems on a daily basis (Emmett et al., 2014a, b).

One key opportunity of particular note for the BESS community was the potential for an EVO-type platform to provide a space to provide a fast and more informed analysis of system change, leading to tools which identify options for immediate, targeted and thus more cost-effective management interventions. This relates to the concept of sudden transitions or tipping points which incur large costs for land and water managers as restoration to previous conditions is difficult and sometimes impossible. Tipping points occur through gradual change in underlying conditions or loss of biodiversity (so-called "slow variables") causing a loss of resilience such that even small perturbations can cause a sudden shift to an alternative state. Critically, this often occurs through a complex interaction of internal biogeochemical, ecological and hydrological processes (Scheffer et al., 2001, 2009; Walker and Salt, 2006). Our current models are a synthesis of our current knowledge and understanding of how land-water systems operate and are therefore hypotheses which can be used to test for unexpected shifts in system behaviour (Dakos et al., 2012). The appearance of novel modes of model failure and/or increasing degradation of model performance even in the presence of improved data quantity and quality may be indicative of fundamental changes occurring in the land–water system itself. If novel integrated data-model systems can reflect and simulate this complex mix of biogeochemical, hydrological and ecological processes to inform new integrated land and water management approaches, in particular, providing early warning of tipping points, there would be significant environmental and economic benefits. The full potential of this approach will only be realised with real-time integration or streaming of observational data with models in a computational platform linking the biogeochemical, hydrological and ecological domains (Emmett et al., 2014a).

The lessons learned and future opportunities from this pilot project are now reflected in a range of community initiatives by NERC, Defra and other organisations. NERC is developing an Environment Research Workbench which is an operational version of the cloud platform explored in the EVO pilot project. A NERC *Environmental Big Data Capital* Call has also enabled large models such as the JULES land-atmosphere model which the EVO tested in its global exemplar to be cloud-enabled to increase access to a broader range of scientists and end-users. Beyond NERC, the lessons learned from EVO are also informing initiatives such as the National Science Foundation's *EarthCube* geoscience project and have also been taken up by a number of end-user organisations in the UK.

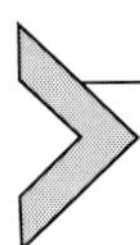

6. UNDERSTANDING AND INTERPRETING BIG DATA THROUGH VISUALISATION

While there are many new statistical approaches being developed for the analysis of Big Data, one of the challenges comes well before this stage, in the generation of hypotheses based on initial observations of interrelationships in the data. Long-term data sets with many different measurements present a significant organisational challenge, and large numbers of traditional simple, static charts can make initial understanding of the data difficult. Indeed, the sheer volume of data and charts may result in key relationships being missed. This can be a special challenge where data are generated from long-term time-series measurements across a large array of different instruments, as is increasingly common in ecological studies. Interactive visualisations of the data provide a means of bringing these disparate sources of data together in new ways, which can then be considered by the researcher using their background knowledge and understanding. One such system for long-term ecological data is the Web-based visual analysis tool Ecological Distributions and Trends Explorer (EcoDATE) (Pham et al., 2013).[11] The tool generates multiple chart views and interaction features, that also support collaboration among multiple users, and is therefore especially well-suited to multi-site consortia projects, where the expertise of different disciplines may be required to understand emerging relationships in the data.

Big Data also present a challenge in relation to communication and impact. Conveying a key message from the data themselves or from the analyses of these data to stakeholders and policy-makers can be very difficult. For ecological data, the traditional depiction of data or results through graphs is too restrictive, and can limit the information conveyed, missing key facts and sometimes resulting in miscommunication or bias (McInerny et al., 2014). In ecosystem ecology, one of the key challenges is conveying the complexity of interactions among species. The presence of an interaction

[11] http://purl.oclc.org/ecodate.

is the most fundamental element of data, but the nature and directionality of these interactions is also important for ecosystem functioning and services, and other measures such as uncertainty and the resilience of an interaction to external pressures, whether natural or anthropogenic, may also be important. Traditional food web diagrams can only convey a fraction of this total information and, for diverse ecosystems, quickly become messy and uninformative. However, new methods of visualisation offer opportunities to begin to convey the important interactions in ecosystems much more effectively and for a more diverse range of audiences (McInerny et al., 2014). In visualisation, as in other realms of ecosystem science, one of the keys to progress in this area is interdisciplinarity. For example, in the business community new visualisation methods are being developed to understand the interrelationships between firms (Basole et al., 2013). Just as these new methods can be used to understand business decisions and opportunities, they also have the potential to be used to understand functional complementarities between different species in ecosystems, both in terms of interactions and functionality or service provision, and could also provide insight into the possible consequences of species loss for ecosystem structure and functioning.

The increasing recognition of the centrality of ecosystems and the environment in policy development across a range of sectors is a welcome development, but it places a considerable burden on the ecosystem community, to be able to respond in an informed and informative manner. New policy developments such as IPBES (McInerny et al., 2014) and progress towards integrative goals such as the Millennium Development Goals and the post-2015 development agenda will require ecosystem data interrogation and interpretation at a range of levels. The ability of the ecosystem science community to generate large quantities of data is increasing rapidly with technology, but the ability of the community to interrogate and communicate the meaning of these data is struggling to keep pace. Information visualisation is an emerging field that could provide ecosystem scientists with some of the tools required to make necessary advances in scientific understanding and communication (Vohland et al., 2011), and ensure that the ecosystem science community can meet the evolving policy demands in this area. New visualisation approaches are currently being developed through the BESS-funded Tansley workshops.[12]

[12] http://www.futureearth.info/blog/2014-jan-27/data-visualization-science-next-frontier.

7. MEETING THE BIG DATA CHALLENGES IN ECOSYSTEM ECOLOGY

It is clear from the foregoing that there is considerable research effort in the UK alone on the science needed to holistically manage natural systems using an Ecosystems Approach based on rigorous ecosystem data and models. The UK Research Councils portfolio in this general area over the past 6 years has included not only the BESS and EVO programmes, but also the *Valuing Nature Network (VNN), Macronutrients Cycles, Ecosystem Services for Poverty Alleviation (ESPA), Insect Pollinator Initiative, Human-modified Tropical Forests, Marine Ecosystems* programmes, and has also contributed to both phases of the *National Ecosystem Assessment*, representing an investment of £70–80M. All these programmes are multi-disciplinary, cover a range of scales and generate huge amounts of data relevant to an Ecosystem Approach, a Big Data challenge. While there are requirements to deposit data within authorised data centres, these are often discipline-specific (see above) and cross-talk between them will be a challenge. In addition, many of these programmes would be informed by historical data held in RCUK data centres and archives listed in Section 3, as well as the plethora of sources shown in Fig. 2.7.

All four of the BESS consortia have scale-based frameworks for data collection and analysis that provide a logical and sensible basis for data integration. However, integrating data across scales is a non-trivial problem (Kemp et al., 2001, Raffaelli and White, 2013). It is vital for researchers to recognise that there are different components to "scale" of grain, lag and extent (defined in Section 1; Wu and Li, 2006a) and to unambiguously identify which component they are dealing with, both within their own sampling hierarchies and when using historical archived data collected at particular scales. Ecological, social and economic processes operate at particular scales of time and space and all three dimensions affect our interpretation of those processes. Indeed, processes are expected to differ at different scales, a real issue when blindly mining Big Data sampled at undefined scales, at a diversity of scales or re-sampled from modelled surfaces. Lack of clarity as to which dimension was varied in a sampling programme could lead to misinterpretations about scale relations within Big Data analyses.

There is also confusion when the term "scale" is used inappropriately (O'Neill and King, 1998): scale has physical dimensions and units of measurement (e.g. kilometres or years), while landscapes, fields and lakes are levels of organisation. Thus, terms like field-scale or stream-scale really have little meaning, although they, and terms like them, are in common usage. It is easy to see how combining "field-scale" data across different studies could therefore lead to all kinds of difficulties for Big Data analyses and researchers need to ensure that the appropriate units of measurement or physical dimensions are reported.

Guidance on understanding how estimates of processes measured for all these dimensions change in upscaling or downscaling is provided by Peterson and Parker (1998), Gardner et al. (2001) and Wu and Li (2006b). They distinguish two basic approaches: similarity-based scaling based on the principle of self-similarity which reduces complex systems to simple mathematical functions (Bloschl and Sivapalan, 1995), and dynamic model-based scaling where models are built to simulate the mechanisms of interest, the parameters of which are modified for different scales (Schneider, 1998; Wu and Li, 2006b). The utility of either approach lies in their ability to detect scaling thresholds where processes change fundamentally.

A second feature of the BESS consortia, also common to other ecosystem-level programmes, is the diversity of types of data collected, especially the challenge of combining qualitative social data with quantitative information from the physical and natural sciences. Many kinds of social science data are easily quantified in specific units, such as farm gate prices, incomes, travel times/distances to markets or recreational sites and employment, or on Likert (e.g. 0–5) scales, such as tranquillity,[13] happiness[14] and the broad range of activities on the MENE (Monitor of Enjoyment in the Natural Environment) database.[15] The pictures that emerge from analyses of such complex data sets can have far-reaching policy implications (e.g. Bateman et al., 2003, 2013), but one needs to be cautious of "The Streetlight Effect" cited at the start of this review. Much social information relevant to our understanding of coupled social-ecological systems, such as indigenous knowledge, cultural meaning and history and related values, is difficult to incorporate in such analyses and therefore usually omitted altogether. These difficulties stem from mismatches in spatial and temporal scales of data collection, a lack of connection between people and landscapes and poor involvement of the social science community in ecosystem-level questions resulting in limited socio-spatial layers for mapping (McLain et al., 2013). In addition, many of the hard-to-capture values have fuzzy spatial

[13] http://www.cpre.org.uk/resources/countryside/tranquil-places/.
[14] http://www.mappiness.org.uk/maps/.
[15] http://www.naturalengland.org.uk/ourwork/evidence/mene.aspx.

and temporal boundaries, creating georeferencing issues. McLain et al. (2013) review a number of developments in the context of environmental planning which may also be helpful for Big Data analyses.

The volume and complexity of Big Data means that ecosystem ecologists also need to revisit the traditional ways in which they generate hypotheses and interrogate data. Multiple variables and complexities in cause and effect, often transcending environmental, social and economic divides and frequently characterised by non-linear relationships and feedbacks, pose considerable challenges to understanding which to date we have tended to ignore or simply not comprehend. New methods for visualising relationships within Big Data have the potential to provide new insights into the functioning and resilience of ecosystems, as well as the types of research questions that we formulate. These methods are likely to lead in turn to the wider application and development of new statistical approaches to characterise complex systems.

Finally, it should be realised that the tension depicted in "The Streetlight Effect" is due to the rapid pace with which novel ideas for analysis and synthesis have emerged as part of the digital revolution, compared to the much slower pace at which data collection has proceeded within much of ecosystem ecology. The resulting mismatch between what we would really love to have as data and what we have to make use of in reality is elegantly captured by the following anecdote:

> *A flashy sports car pulls up adjacent to a field in deepest and remotest rural Norfolk, and the driver leans out to ask a farm labourer: "I say, my man, do you know the way to Baconsthorpe?" The labourer thinks for a minute and replies: "Yes, but if I were you, I wouldn't start from here".*

Unfortunately, whether we like it or not, it is "here" from which we have to start when it comes to Big Data. The challenge is to find novel ways to make best use of the data we have and what we can measure.

ACKNOWLEDGEMENTS

All the authors are extremely grateful to the Natural Environment Research Council for their support of the BESS and EVOp programmes. Two anonymous reviewers greatly improved the chapter.

REFERENCES

Basole, R.C., Clear, T., Mengdie, Hu., Mehrotra, H., Stasko, J., 2013. Understanding interfirm relationships in business ecosystems with interactive visualization. IEEE Trans. Vis. Comput. Graph. 19, 2526–2535.

Bateman, I.J., Lovett, A.A., Brainard, J.S., 2003. Applied Environmental Economics. A GIS Approach to Cost-Benefit Analysis. Cambridge University Press, Cambridge.

Bateman, I., Harwood, A., Mace, G., Watson, S.R., Abson, D.J., Andrews, B., Binner, A., Crowe, A., Day, B., Dugdale, S., Fezzi, C., Foden, J., Hadley, D., Haines-Young, R., Hulme, M., Kontoleon, A., Lovett, A., Munday, P., Pascual, U., Paterson, J., Perino, G., Sen, A., Siriwardena, G., Van Soest, D., Termansen, M., 2013. Bringing ecosystem services into economic decision-making: land Use in the united kingdom. Science 341, 45–50.

Bloschl, G., Sivapalan, M., 1995. Scale issues in hydrological modelling: a review. Hydrol. Process. 9, 251–290.

Boffey, P.M., 1976. International biological programme: was it worth the cost and effort? Science 193, 866–868.

Collins, K.J., Weiner, J.S., 1977. Human Adaptability. A History and Compendium of Research. Taylor & Francis Ltd./St. Martin's Press Inc., London/New York. 356 p.

Convention on Biological Diversity, 2009. http:/www.cbd.int/ecosystem/ (sourced January 2009).

Dakos, V., Carpenter, S.R., Brock, W.A., Ellison, A.M., Guttal, V., Ives, A.R., Kéfi, S., Livina, V., Seekell, D.A., van Nes, E.H., Scheffer, M., 2012. Methods for detecting early warnings of critical transitions in time series illustrated using simulated ecological data. PLoS One 7, 1–20.

Eigenbrod, F., Armsworth, P.R., Anderson, B.J., Heinemeyer, A., Gillings, S., Roy, D.B., et al., 2010. The impact of proxy-based methods on mapping the distribution of ecosystem services. J. Appl. Ecol. 47, 377–385.

Emmett, B.A., Gurney, R.J., McDonald, A.T., Blair, G., Buytaert, W., Freer, J., Haygarth, P., Rees, G.H., Tetzlaff, D., Afgan, E., Ball, L. A., Bevan, K., Bick, M., Bloomfield, J. B., Brewer, P., Delve, J., Donaldson-Selby, G., Johnes, P.J., El-Khatib, Y., Field, D., Gemmell, A.L., Ghimire, S., Greene, S., Huntingford, C., Lewis, J., Mackay, E., Macklin, M.V., Macleod, K., Marshall, K, Odoni, N., Percy, B.J., Quinn, P.F., Reaney, S., Stutter, M., Surajbali, B., Thomas, N.R, Vitolo, C., Watson, H, Williams, B.P., Wilkinson, M., Zelazowski, P. 2014a. NERC Environmental Virtual Observatory Pilot. Final Report. November 2013, Grant No. NE/I002200/1.

Emmett, B.A., Gurney, R.J., McDonald, A.T., Ball, L.A., Beven, K., Bicak, M., Blair, G., Bloomfield, J., Buytaert, W., Delve, J., Elkhatib, Y., Field, D., Freer, J., Gemmel, A., Greene, S., Haygarth, P., Huntingford, C., Johnes, P., Mackay, E., Macklin, M., Macleod, K., Odoni, N., Percy, B., Quinn, P., Reaney, S., Rees, G., Stutter, M., Surajbali, B., Tetzlaff, N., Vitolo, C., Wilkinson, M., Williams, B., Zelazowski, P., 2014a. Heads in the clouds. Int. Innov. 141, 81–85.

Freedmand, D.H., 2010. Why scientific studies are so often wrong: the streetlight effect. Discover, July/August issue.

Gardner, R.H., Kemp, W.M., Kennedy, V.S., Petersen, J.E., 2001. Scaling Relations in Experimental Ecology. Columbia University Press, New York.

Gómez-Baggethun, E., de Groot, Rudolf, Lomas, Pedro L., Montes, Carlos, 2010. The history of ecosystem services in economic theory and practice: from early notions to markets and payment schemes. Ecol. Econ. 6, 1209–1218.

Gray, J., 2007. Talk given by Jim Gray to the NRC-CSTB in Mountain View, CA, on January 11, 2007. http://research.microsoft.com/en-us/um/people/gray/talks/NRC-CSTB_eScience.ppt.

Hampton, S., Strasser, C., Tewksbury, J., Gram, W., Budden, A., Batcheller, A., Duke, C., Porter, J., 2013. Big data and the future of ecology. Front. Ecol. Environ. 11, 156–162.

Hey, T., Tansley, S., Tolle, K. (Eds.), 2009. The Fourth Paradigm Data-Intensive Scientific Discovery. Microsoft Research, Redmond, WA.

Howe, D., Costanzo, M., Fey, P., Gojobori, T., Hannick, L., Hide, W., et al., 2008. Big data: the future of biocuration. Nature 455, 47–50.
Jones, K.E., Paramor, O.A.L., 2010. Inter-disciplinarity in ecosystems research: developing social robustness in environmental science. In: Frid, C.L.J., Raffaelli, D.J. (Eds.), Ecosystem Ecology: A New Synthesis. In: BES Ecological Review Series, Cambridge University Press, Cambridge, UK.
Kareiva, P., Tallis, H., Ricketts, T.H., Daily, G.C., Polasky, S., 2011. Natural Capital: Theory and Practice of Mapping Ecosystem Services. Oxford University Press, Oxford.
Kemp, W.M., Petersen, J.E., Gardner, R.H., 2001. Scale-dependence and the problem of extrapolation: implications for experimental and natural coastal ecosystems. In: Gardner, R.H., Kemp, W.M., Kennedy, V.S., Petersen, J.E. (Eds.), Scaling Relations in Experimental Ecology. Columbia University Press, New York, pp. 3–60.
Lowe, P., Phillipson, J., 2006. Reflexive interdisciplinary research: the making of a research programme on the rural economy and land use. J. Agric. Econ. 57, 165–184.
Lowe, P., Whitman, G., Phillipson, J., 2009. Ecology and the social sciences. J. Appl. Ecol. 46, 297–305.
McInerny, G.J., Chen, M., Freeman, R., Gavaghan, D., Meyer, M., Rowland, F., Spiegelhalter, D.J., Stefaner, M., Tessarolo, G., Hortal, J., 2014. Information visualisation for science and policy: engaging users and avoiding bias. Trends Ecol. Evol. 29, 148–157.
McLain, R., Poe, M., Biedenweg, K., Cerveny, L., Besser, D., Blahna, D., 2013. Making sense of human ecology mapping: an overview of approaches to integrating socio-spatial data into environmental planning. Hum. Ecol. 41, 651–665. http://dx.doi.org/10.1007/s10745-013-9573-0.
Michener, M., Jones, M., 2012. Ecoinformatics: supporting ecology as a data-intensive science. Trends Ecol. Evol. 27, 85–93.
Millennium Ecosystem Assessment, 2005. Ecosystems and Human Well-being: Synthesis. Island Press, Washington, DC.
National Ecosystem Assessment, 2014. The UK National Ecosystem Assessment: Synthesis of the Key Findings. UNEP-WCMC, LWEC, UK.
Natural England, 2012. Health and Natural Environments—An Evidence Based Information Pack. Natural England, Peterborough, UK, p. 12.
Nelson, E., Mendoza, G., Regetz, J., Polasky, S., Tallis, H., Cameron, D., et al., 2009. Modeling multiple ecosystem services, biodiversity conservation, commodity production, and tradeoffs at landscape scales. Front. Ecol. Environ. 7, 4–11.
NERC. 2014. Environmental Virtual Observatory Pilot. Final report. NERC, Swindon, 80 p.
O'Neill, R.V., King, A.W., 1998. Homage to St Michael: or why are there so many books on scale? In: Peterson, D.L., Parker, V.T. (Eds.), Ecological Scale. Theory and Applications. Columbia University Press, New York, pp. 3–16.
Peterson, D.L., Parker, V.T., 1998. Ecological Scale. Theory and Applications. Columbia University Press, New York.
Pham, T., Jones, J., Metoyer, R., Swanson, F., Pabst, R., 2013. Interactive visual analysis promotes exploration of long-term ecological data. Ecosphere 4, art112.http://dx.doi.org/10.1890/ES13-00121.1.
Raffaelli, D., 2000. Trends in research on shallow water food webs. J. Exp. Mar. Bio. Ecol. 250, 223–232.
Raffaelli, D.G., Frid, C.L.J., 2010. Ecosystem Ecology: A New Synthesis. Cambridge University Press, Cambridge.
Raffaelli, D., White, P.C.L., 2013. Ecosystems and their services in a changing world: an ecological perspective. Adv. Ecol. Res. 48, 1–70.

Scheffer, M., Carpenter, S., Foley, J.A., Folke, C., Walker, B., 2001. Catastrophic shifts in ecosystems. Nature 413, 591–596.

Scheffer, M., Bascompte, J., Brock, W.A., Brovkin, V., Carpenter, S.R., Dakos, V., Held, H., van Nes, E.H., Rietkerk, M., Sugihara, G., 2009. Early-warning signals for critical transitions. Nature 461, 53–59.

Schneider, D.C., 1998. Applied scaling theory. In: Peterson, D.L., Parker, V.T. (Eds.), Ecological Scale. Theory and Applications. Columbia University Press, New York, pp. 253–270.

Tansley, A.G., 1935. The use and abuse of vegetational concepts and terms. Ecology 16, 284–307.

The Royal Society, 2012. Science as an Open Enterprise. The Royal Society Policy Centre.

UK National Ecosystem Assessment, 2011. The UK National Ecosystem Assessment: Synthesis of the Key Findings. UNEP-WCMC, Cambridge.

Vohland, K., Mlambo, M.C., Horta, L.D., Jonsson, B., Paulsch, A., Martinez, S.I., 2011. How to ensure a credible and efficient IPBES? Environ. Sci. Policy 14, 1188–1194.

Walker, B., Salt, D., 2006. Resilience Thinking. Sustaining Ecosystems and People in a Changing World. Island Press, Washington, p. 174.

Weiner, J.S., Lourie, J.A., 1969. Human Biology: A guide to Field Methods. Blackwells Scientific Publications, Oxford, UK.

White, P.C.L., Cinderby, S., Raffaelli, D., de Bruin, A., Holt, A., Huby, M., 2009. Enhancing the effectiveness of policy-relevant integrative research in rural areas. Area 41, 414–424.

Willis, A.J., 1997. The ecosystem: an evolving concept viewed historically. Funct. Ecol. 11, 268–271.

Worthington, E.B., 1969. The international biological programme. Nature 208, 223–226.

Worthington, E.B., 1975. The Evolution of the IBP. Cambridge University Press, Cambridge, UK, p. 268.

Worthington, E.B., 1983. The Ecological Century. A Personal Appraisal. Calrendon Press, Oxford, p. 199.

Worthington, E.B., Fogg, G.E., Waddington, C.H., Clymo, R.S., Newbould, P.J., Holdgate, M.W., Clarke, Cyril, 1976. General discussion. Philos. Trans. R. Soc. Lond. B 274, 499–507.

Wu, J., Li, H., 2006a. Concepts of scale and scaling. In: Wu, J., Jones, K.B., Li, H., Loucks, O.L. (Eds.), Scaling and Uncertainty Analysis in Ecology. Methods and Applications. Springer, Dordecht, pp. 3–16.

Wu, J., Li, H., 2006b. Perspectives and methods of scaling. In: Wu, J., Jones, K.B., Li, H., Loucks, O.L. (Eds.), Scaling and Uncertainty Analysis in Ecology. Methods and Applications. Springer, Dordecht, pp. 17–44.

CHAPTER THREE

DNA Metabarcoding Meets Experimental Ecotoxicology: Advancing Knowledge on the Ecological Effects of Copper in Freshwater Ecosystems

Stephanie Gardham[*,‡,1], **Grant C. Hose**[†], **Sarah Stephenson**[‡], **Anthony A. Chariton**[‡]

[*]Department of Environment and Geography, Macquarie University, Sydney, New South Wales, Australia
[†]Department of Biological Sciences, Macquarie University, Sydney, New South Wales, Australia
[‡]CSIRO Oceans and Atmosphere, Lucas Heights, New South Wales, Australia
[1]Corresponding author: e-mail address: stephgardham@hotmail.com

Contents

Abstract

DNA (DNA) metabarcoding is a molecular tool that may revolutionise the way in which biological communities are assessed. The tool has the potential to allow a much larger proportion of the biological community to be identified more reliably and rapidly than by current methods of analysis, including meio- and microbiota that would otherwise

Advances in Ecological Research, Volume 51
ISSN 0065-2504
http://dx.doi.org/10.1016/B978-0-08-099970-8.00007-5

be missed. Here, DNA metabarcoding was performed to assess the effects of copper on the establishment of a benthic eukaryote community within a series of environmentally relevant copper-contaminated, freshwater mesocosms. The organisms present from the micro- to macroscale were characterised using this method. While taxonomic richness of the eukaryote community increased in control and low copper treatments (<200 mg/kg dry wt particulate copper; <5 μg/L pore water copper), it remained constant or declined over time in higher copper treatments (>400 mg/kg dry wt; >18 μg/L). The response observed in the composition of the benthic eukaryote community was more subtle, with significant differences apparent between all treatments during the initial establishment of the community, even the control (4.6 mg/kg dry wt; 1.5 μg/L) and very low (71 mg/kg dry wt; 2.8 μg/L) copper-contaminated treatments. This response was much more sensitive than that observed by a traditional analysis of the macroinvertebrate community over the same time period. All the taxa identified to be sensitive to copper were meio- and microbiota including Chlorophyta, Nematoda, Bacillariophyta and Fungi. This study demonstrates the potential power of DNA metabarcoding for ecotoxicological studies and emphasises the need to incorporate and meio- and microbiota into bioassessment processes.

1. INTRODUCTION

Copper is a key metal contaminant of aquatic environments. Sources of the metal include urban waste water, industrial and mine effluents, agricultural runoff and atmospheric deposition (Serra and Guasch, 2009). Like other metals, once copper enters aquatic environments it becomes adsorbed to sediment particles (Chon et al., 2012). Consequently, concentrations of copper in sediments often far exceed those in overlying waters and, as a result, sediments become sources of copper, causing adverse effects on benthic biota in particular (Chapman et al., 1998; Harrahy and Clements, 1997).

There has been extensive research into the effects of copper on aquatic biota; however, this has focused on how the metal affects individual species; 96% of the records held within the U.S. EPA Ecotoxicology database on freshwater copper toxicity were derived from laboratory studies, of which only 2% assessed end points involving more than one species (USEPA, 2013). Laboratory studies are invaluable to identify the specific mechanisms through which a contaminant, like copper, can exert an effect on one or few organism(s) (Clements, 2000). They can be used to identify how contaminants interact at target sites in organisms, which may be linked to physiological (e.g. respiration) and individual (e.g. mortality) level responses (Clements, 2000; Maltby, 1999). However, laboratory studies often have questionable ecological relevance as they may not use species representative of those in the field, are often carried out over short exposure times under very controlled conditions, and to extrapolate data to the field it must be assumed that individual organisms function as discrete units and are therefore not affected by intra- or interspecific interactions (Fleeger et al., 2003; Mayer-Pinto et al., 2010; Preston, 2002).

Studies at the community level are often carried out under field or near-field conditions and can be more ecologically relevant than laboratory studies (Clements, 2000). Fewer assessments of the effects of contaminants, however, have been carried out at the community level because they are expensive, time consuming and labour intensive. For copper, assessments at the community level have generally focused on the macroinvertebrate community (e.g. Clements et al., 1988; Shaw and Manning, 1996), periphyton community (e.g. Kaufman, 1982; Serra et al., 2009) and plankton community (e.g. Havens, 1994; Le Jeune et al., 2006; Swartzman et al., 1990). Some studies have considered multiple fractions of the overall community; for example, recently, the effects of copper on the phytoplankton, periphyton, macroinvertebrate and macrophyte communities in freshwater lotic mesocosms was assessed (Roussel et al., 2007, 2008). Nevertheless, the effects of copper on the community composition of meio- and microbiota in aquatic ecosystems is poorly studied in comparison to macrofauna because of the relative difficulty in sampling and identification (Kennedy and Jacoby, 1999). This is despite evidence that meio- and microbiota are more sensitive than macrofauna to metals (Kennedy and Jacoby, 1999). Meio- and microbiota are also essential components of the aquatic community, with key roles in many globally important biogeochemical cycles, for example, primary production and nitrogen assimilation (Gadd and Raven, 2010). Thus, research into the effects of copper (and other contaminants) on the community composition of meio- and microbiota is needed, but to date the potential for this has been limited.

A new tool has emerged from molecular science in recent years through which a far larger proportion of the biological community present can be identified compared to previous techniques of analysis (Chariton et al., 2010). In DNA metabarcoding, extracts of DNA from environmental samples can be analysed to identify multiple taxa at the same time (Yoccoz, 2012). Amplicons (informative fragments of DNA) obtained via PCR of the DNA extract are sequenced by Next Generation Sequencing technology. The sequences are compared to online or custom reference libraries, e.g. GenBank and SILVA, to assign taxonomic names to biota present in the environmental samples (Shokralla et al., 2012). The costs of sequencing amplicons has decreased exponentially year on year, and as a result, DNA metabarcoding has the potential to not only assess a greater proportion of the biological community more reliably and rapidly than ever before, but also be a cost-effective form of assessment (Bohmann et al., 2014).

DNA metabarcoding has already led to a greater understanding of biodiversity. In particular, the tool has allowed the identification of rare or difficult-to-sample organisms, generally meio- and microbiota, which otherwise would not be considered during assessments (Bohmann et al., 2014). It has been used to assess changes in alveolate turnover in lakes (Medinger et al., 2010), root-associated fungal communities in glacier forelands (Blaalid et al., 2012) and eukaryotic biodiversity on floodplains (Baldwin et al., 2013) among many other studies (e.g. Chariton et al., 2014; Creer et al., 2010). These studies have demonstrated that, through DNA metabarcoding, extensive data from a standard environmental sample can be extracted, which are ecologically relevant at the whole ecosystem scale (Baird and Hajibabaei, 2012; Chariton et al., 2010).

In this study, the effects of copper on the establishment of benthic eukaryote communities in a series of outdoor mesocosms over 1.5 years were assessed. The primary hypothesis of the present study was that, with an increase in copper, a decrease in the richness of the benthic eukaryote community and change in its composition would be observed. The paper builds upon an analysis of the establishment of the macroinvertebrate community within the mesocosms over the same time period (Gardham et al., 2014a). A salient feature of this research is the different ecological trends observed using DNA metabarcoding compared to the analysis of the macroinvertebrate community.

2. METHODS

2.1. Study design

Twenty pond mesocosms were established on the Macquarie University campus (33.76946°S, 151.11496°E), in northern Sydney, NSW, Australia. Full details of the infrastructure are described in Gardham et al (2014b). Briefly, each mesocosm had a volume of 1500 L and was sunk into the ground. The mesocosms were shaded by 70% shade-cloth, aerated, and a consistent water depth was maintained via a drainage system and rainwater inputs.

Sediments were spiked *in situ* with copper and allowed 2 months to equilibrate and create environmentally relevant copper-contaminated sediment prior to opening the mesocosms up to colonisation on 1 November 2010 (day 0) (Gardham et al., 2014b). The experimental design consisted of a control (C) and four sediment copper concentrations (very low (VL), low (L), high (H) and very high (VH)), each with four replicates (Table 3.1); see Gardham et al. (2014b) for detailed information on the treatment design. On each sampling occasion, measurements of the concentrations of copper in the sediments, pore waters and overlying waters and several water quality parameters (conductivity, pH, turbidity, dissolved oxygen and redox potential) were made. A detailed analysis of these parameters showed that the primary abiotic difference between the treatments was copper (Gardham et al., 2014a).

2.2. Sampling strategy

Sediment samples were taken from each mesocosm four times during the first spring/summer season (15, 31, 135 and 161 days) and three times during the second spring/summer season (365, 407 and 497 days). Three cores (6 cm diameter) were randomly collected from the surficial layer of the sediment (top 1–2 cm) using a plastic corer. They were combined in a plastic sample jar, which was then capped and inverted six times to mix the cores together. A 50-mL sterile tube was filled with a sub-sample of the sediment and stored on ice in the field. On return to the laboratory (within 2 h), the samples were frozen at −80 °C until processing.

Corers and sample mixing jars were sterilised by soaking in 15% bleach solution for 24 h followed by a thorough rinse with tap water and then UV sterilisation (10 min) prior to each sampling occasion. The 50-mL sterile

Table 3.1 Mean ± SEM ($n = 4$) copper concentrations in the sediment, pore waters and overlying waters by treatment

Treatment	Sediment copper (mg/kg dry wt)	Pore water copper (μg/L)	Overlying water copper (μg/L)
Control	4.6 ± 0.7	1.5 ± 0.2	1.3 ± 0.3
Very low	71 ± 4.6	2.8 ± 0.2	2.2 ± 0.2
Low	99 ± 7.1	4.6 ± 1.2	3.8 ± 0.5
High	410 ± 41	18 ± 3.1	7.8 ± 0.6
Very high	711 ± 46	30 ± 2.5	11 ± 0.9

collection tubes were also UV sterilised to ensure sterility. Separate sampling equipment was used for each mesocosm to prevent cross-contamination.

2.3. Molecular analysis

This study utilised the 18S rRNA gene as a marker. This is consistent with most eukaryote surveys (Tang et al., 2012). The taxonomic level that can be assigned based on molecular analysis is dependent on the length of the sequenced gene, whether the amplified target region has previously been sequenced for the relevant taxa (at an appropriate quality), and the homology of that gene to the annotated sequence (Chariton et al., 2010). There are large quantities of appropriate data available for the 18S rRNA gene because it is often sequenced when investigating divergence rates and establishing phylogenetic information for individual species (Chariton et al., 2010).

DNA was extracted and purified from a 10 g sub-sample of each homogenised surficial sediment sample using PowerMax Soil DNA Isolation Kits (MO BIO Laboratories Inc., Carlsbad, CA). After performing a gradient PCR to establish the optimal annealing temperature (58 °C) and template DNA dilution (1:6), PCR was performed on each of the DNA extracts using the primers All18SF (5′-TGGT GCATGGCCGTTCTTAGT-3′) and All18SR (5′-CATCTAAGG GCATCACAGACC-3′). This primer pair has high coverage, theoretically able to amplify 18S target sequences from >85% of species across all Kingdoms (Baldwin et al., 2013), and targets between 200 and 500 bp in the 3′ region of the gene encoding 18S rRNA. PCR amplifications were performed in 50 μl volumes with Phusion High Fidelity (HF) Master Mix (Thermo Fisher Scientific Inc., Pittsburgh, PA) in the buffer supplied and 0.5 μM All18SF, 0.5 μM All18SR, nuclease-free water, DMSO and 2 μL of template DNA. Amplification was performed in an Eppendorf thermocycler (Eppendorf, Hamburg, Germany) with the following cycling: one cycle of 98 °C for 1 min, then 35 cycles of 98 °C for 10 s, 58 °C for 20 s and 72 °C for 30 s, followed by one cycle of 72 °C for 5 min.

All PCRs were run with a positive artificial assemblage control and a negative water control to detect contamination and to test bioinformatics parameters. PCR integrity and amplicon bp size were assessed using MultiNA (Shimadzhu, Japan) digital gel analysis. The 18S amplicons were purified using an AMPure XP purification kit (Agencourt, Beverly, MA) and quantified by NanoDrop spectrophotometry (NanoDrop 2000, Thermo Fisher Scientific Inc.). Custom-designed fusion primers were ligated to the 18S amplicons to incorporate 454 pyrosequencing adapter A and B sequences and unique 10 base barcode sequences were attached to each sample for multiplexing to the ends of the 18S target sequences, as required for 454 pyrosequencing. This was achieved using a four cycle ligation PCR, performed in 50 μL volumes with Phusion HF Master Mix and 15 μL template DNA (which comprised 200 ng of the 18S amplicons—ensuring equimolar concentrations and even ligation of fusion primers across amplicons within samples). The PCR cycle was: one cycle of 98 °C for 2 min, and then four cycles of 98 °C for 30 s, 54 °C for 1 min and 72 °C for 1 min, followed by one cycle of 72 °C for 3 min. These amplicons were purified, quantified and pooled to achieve a target of 300 ng/μL then purified again using QIAquick PCR purification kit (Qiagen, Valencia, CA). Pyrosequencing of the pooled 18S amplicon libraries was performed using a two-region large gasket format on a Roche 454 GS FLX sequencer with Titanium chemistry at the Australian Genome Research Facility (Brisbane, Queensland, Australia).

2.4. Sequence analysis

Demultiplexing and the removal of potential PCR artefacts, sequencing errors and chimeric sequences, was performed using the default parameters of the bioinformatics pipeline Amplicon Pyrosequence Denoising Program (APDP) (Morgan et al., 2013). APDP is a novel approach for demultiplexing and denoising which uses information on the abundance distribution of similar sequences across independent samples, as well as the frequency and diversity of sequences within individual samples, to accurately reflect the true sequence richness of pyrosequenced compositional data. Sequences were processed using APDP if they met four initial criteria: (1) an exact match for the forward multiplex identifier (MID) sequence and All18SF primer; (2) an exact match for the reverse MID sequence and All18SR primer; (3) a length >80 bp, including primer; and (4) identifiable combinations of the forward and reverse MIDs as defined by the sequencing strategy.

The data set of sequences that met the initial criteria for APDP was first filtered to remove sequences with ambiguous nucleotides and singletons. The remaining sequences were then binned to their sample of origin and cropped to remove their MIDs and primer sequences. The filtered sequences were assigned to groups based on the accession numbers of the best hit as determined by MEGABLAST against the NCBI non-redundant database. Sequences that did not return a hit were removed. Unique sequences with the most overall reads (excluding singletons) and any other sequence with the most reads in any individual sample were retained. Additional filtering was then performed to remove sequences that were only present in a single PCR reaction and had less than 10 reads. Finally, three way alignments of all possible combinations of sequences were conducted within each sample to identify potential chimeric and erroneous sequences. Upon completion, the remaining unique sequences provided by APDP were considered to be valid, separate and independent molecular operational taxonomic units (MOTUs).

Using the default settings in the bioinformatics package QIIME 1.6 (Caporaso et al., 2010), taxonomic identification for each validated MOTU was inferred against the SILVA database (release 113) and GenBank (www.arb-silva.de/). Additional taxonomic information was obtained using megaBLAST to retrieve the top-scoring hit (total score) for each sequence existing in GenBank as of 10 August 2014. To provide consistency, the taxonomy for all MOTUs identified by GenBank was manually corrected to the nomenclature used by The National Center for Biotechnology Information (http://www.ncbi.nlm.nih.gov/Taxonomy/taxonomyhome.html).

2.5. Statistical analysis

All samples were rarefied to 3796 reads prior to analysis using the R package VEGAN. This number of reads represented the lowest number of reads in any of the samples. The adequacy of the sampling regime in capturing the diversity of the mesocosms system was examined by comparing the observed richness with that predicted using the Chao2 estimator (Chao, 1984). This analysis was done using Primer-E v6 (Primer-E Ltd, Plymouth, UK).

The effect of time and treatment on the mean total MOTU richness per sample was assessed using repeated measures (RM) ANOVAs (between-subjects factor: treatment, within-subjects factor: time) using SPSS v21. If the assumption of sphericity (assessed by Mauchly's test) was not met during the RM ANOVA analyses, the Greenhouse–Geisser correction was used. Differences in MOTU richness among treatments on each sampling occasion were tested using a one-way ANOVA followed by LSD pair-wise comparisons between treatments and controls.

Principal response curves (PRC) analysis (Van den Brink and Ter Braak, 1999) was used to examine the temporal changes between controls and treatments. PRC is a multivariate method of analysis that highlights differences in community composition between treatments and the control at each time point. The species weights indicate the effect at the species level; the greater the weight (positive or negative), the more closely the abundance of that taxon follows the patterns in the PRC ordination. The significance of the PRC was tested using Monte Carlo permutation tests. The PRC analysis was based on Euclidean distance and $\log_{10}(x+1)$ transformed data and was done using Canoco version 5 (Ter Braak and Smilauer, 2012); recommended software settings were used.

Non-metric multidimensional scaling (nMDS) using Bray-Curtis dissimilarity and $\log_{10}(x+1)$ transformation of data was used to further investigate differences in community composition amongst samples. This analysis was done using Primer-E v6 (Primer-E Ltd, Plymouth, UK). Permutational Multivariate Analysis of Variance (PERMANOVA) (Anderson, 2001) was used to further test the observed changes in assemblage structure between treatments and over time. The PERMANOVA analysis was based on the Euclidean distance and $\log_{10}(x+1)$ transformed data to mirror that used for the PRC analysis. The PERMANOVA was done using Primer 6+.

Distanced-based linear modelling (distLM) (McArdle and Anderson, 2001) was performed to examine the relative influence of the sediment, overlying water and pore water copper concentrations and time on community composition. The analysis was based on the Bray–Curtis dissimilarity matrix of the $\log_{10}(x+1)$ transformed biological data and normalised environmental data. The significance level (α) for all statistical tests was 0.05.

We emphasise that there are some caveats associated with the approach used in this study. Most notably, it has been well documented that a number of technical and biological factors constrain the use of quantifiable data from PCR-based techniques such as DNA metabarcoding (Angly et al., 2014; Deagle et al., 2013; Egge et al., 2013). However, the use of such data within a semi-quantitative framework, where the composition of samples is estimated by the proportion of reads corresponding to a particular MOTU, is common place (e.g. Bradford et al., 2013; Willerslev et al., 2014). Using this approach, comparisons are restricted to relative changes in read counts between and not within samples. To ensure that the fundamental ecological patterns observed in this study were not significantly altered by choosing this approach, direct comparisons between $\log_{10}(x+1)$ transformed and presence–absence data were made using nMDS plots. The two approaches showed a strong agreement with the key composition patterns remaining unchanged (Figs. 3.1 and 3.A1).

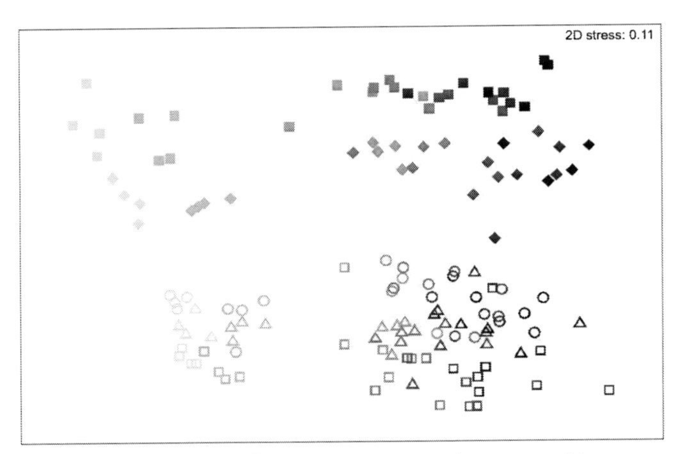

Figure 3.1 nMDS plot showing differences in community composition among samples based on $\log_{10}(x+1)$ data, where points closer together represent a more similar community composition than those further apart. Key: control (open square), very low (open triangle), low (open circle), high (solid diamond), very high (solid square) and gradient of light blue (light grey in the print version) to dark blue (dark grey in the print version) denotes change in time from 15 days through to 497 days.

3. RESULTS

The sequencing run produced over 1.3 million reads and 1,132,453 reads remained after APDP processing. The final rarefied data contained 531,440 reads, representing 2202 MOTUs which were recorded as present 31,567 times across all time points and samples. These data are available in the Commonwealth Scientific and Industrial Research Organisation data access portal at: http://dx.doi.org/10.4225/08/5428AFA7DF112.

By the end of the study, the accumulation of species had reached an asymptote and the sampling regime had collected 97.0% of the estimated taxa richness. The MOTUs were identified across 76 phyla (and other coarse taxonomic levels), 76 classes, 269 orders, 380 families and 439 genera. Microbes dominated the benthic eukaryote communities (Fig. 3.2); the most "MOTU rich" phyla present across all treatments and sampling occasions were Chlorophyta (9% of overall richness), Ciliophora (8%), Ascomycota (7%), Nemertea (6%) and Basidiomycota (6%).

In general terms, the MOTU richness of the C, VL and L treatments increased over time, whereas the richness of the H and VH treatments remained relatively constant or declined over time (Fig. 3.3). This trend

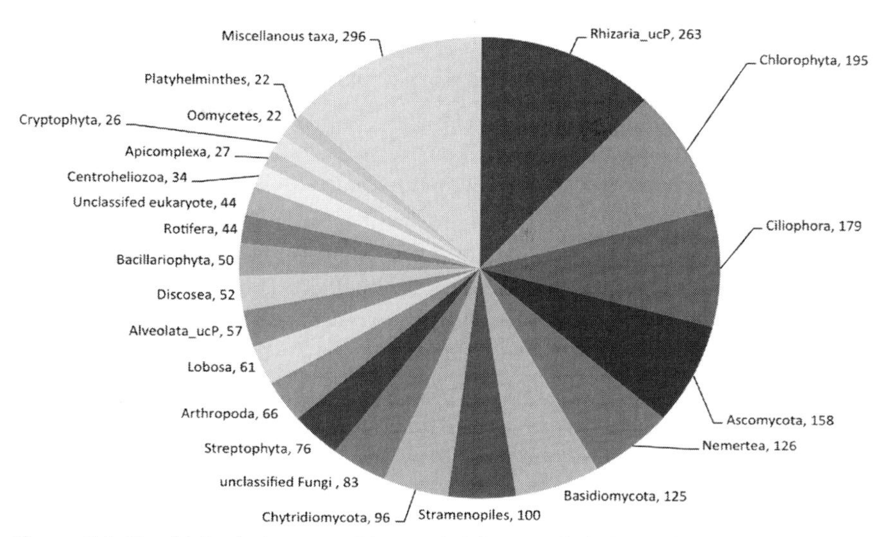

Figure 3.2 The biological composition and richness of phyla (and other coarse taxonomic groups) sequenced from the surficial sediments from all treatments and time points. Data are shown at the level of phylum and higher for ease of interpretation.

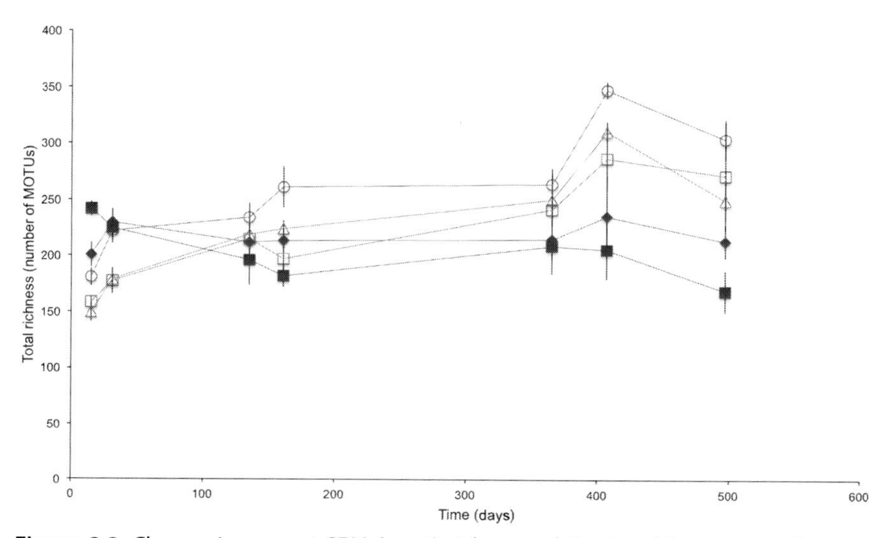

Figure 3.3 Change in mean ± SEM ($n=4$) richness of the benthic communities over time by treatment. Key: control (open square), very low (open triangle), low (open circle), high (solid diamond) and very high (solid square).

is reflected in a significant time × treatment interaction ($p < 0.001$). There were also significant time and treatment main effects ($p < 0.007$). At 15 and 31 days, the mean MOTU richness of the VL treatment was not significantly different to that of the C treatment ($p > 0.072$), but the higher concentration treatments (L, H and VH) had richness significantly greater than that of the controls ($p \leq 0.006$). At 135 and 365 days, there was no difference in richness between the controls and any treatments ($p \geq 0.283$). At 161 days, the mean MOTU richness of the C was significantly lower than that of the L treatment ($p = 0.003$), but not the VL and higher concentration treatments (H and VH) ($p \geq 0.157$). At 407 and 497 days, the mean MOTU richness of the controls was significantly higher than that of the VH treatment ($p \leq 0.031$), but not the other copper treatments (VL, L, H) ($p \geq 0.091$).

The composition of the eukaryote communities was significantly different among treatments on the first sampling occasion (15 days; 2 months after spiking). Sediments for each mesocosm were sourced from the same batch of homogenised soil, so we assume that they each had similar assemblages of meio- and microbiota prior to spiking. While the spiking method successfully excluded invertebrates (Gardham et al., 2014a), it was not possible to exclude meio- and microbiota. The assemblages of meio- and microbiota present at 15 days reflect the development of the community in response to copper as well as a transition from soil (terrestrial) to aquatic sediment assemblages.

The differences in community composition among sampling times described accounted for 16% of the total variance in community composition while 30% of the variance was explained by the differences in community composition due to treatment. The first axis of the PRC was significant ($p = 0.002$) and explained 34.4% of the variance captured by the treatment regime (Fig. 3.4A). The second axis was also significant ($p = 0.005$) but only explained 11.9% of the variance and showed a similar trend as the first PRC axis in which the H and VH treatments together follow a similar trajectory, but one that is different from that of the C, VL and L treatments (Fig. 3.4B).

The first PRC axis highlights the response of MOTUs for which the number of reads deviate relatively consistently in the copper treatments compared to the controls over the course of the study (Fig. 3.4A). The MOTUs that most strongly followed the response shown in the first PRC axis and were positively affected by copper in the H and VH treatments include three chlorophytes (*Scenedesmus* sp., *Paulschulzia* sp. and *Neochlorosarcina*) and one streptophyte (*Cosmarium* sp.), two amoeboid protozoa (*Hartmannella sp.* and *Protosporangium*), one flagellate protozoa

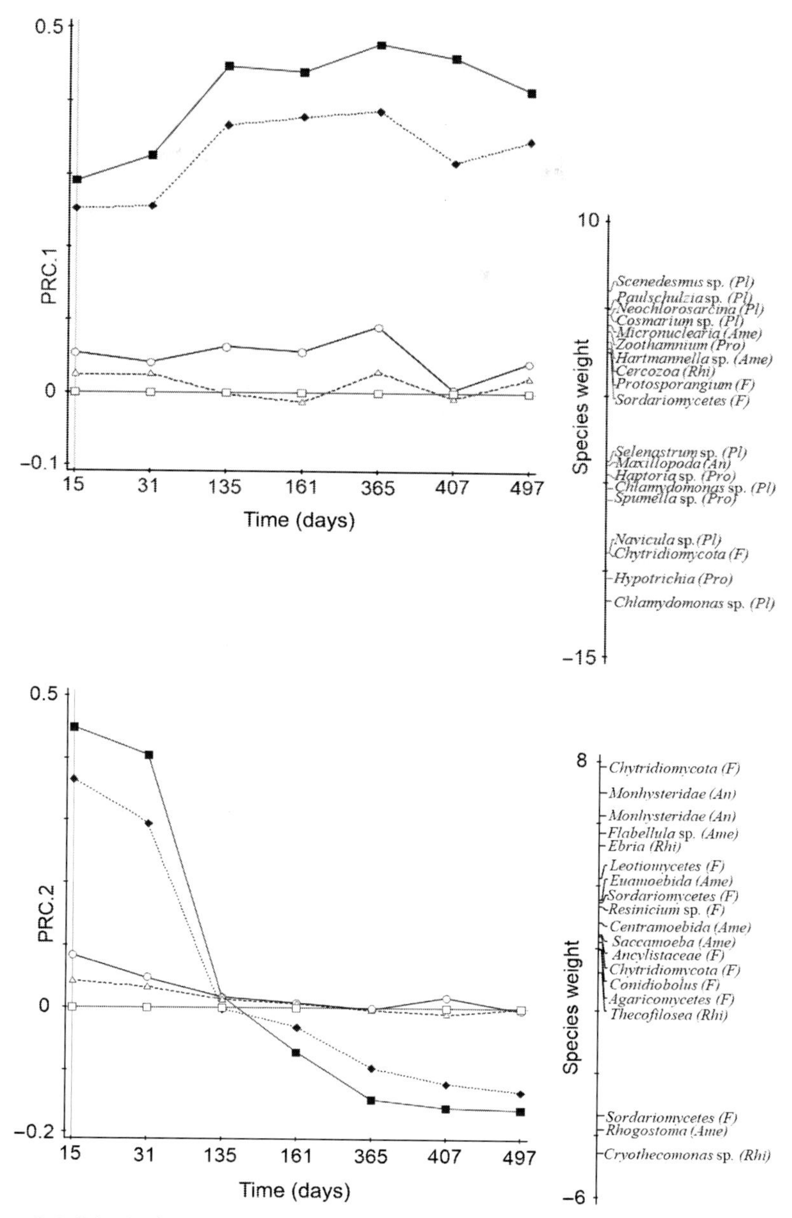

Figure 3.4 Principal response curves axis 1 (A) and axis 2 (B), with species weights showing the effect of copper on establishment of the benthic invertebrate community. Key for each PRC: control (open square), very low (open triangle), low (open circle), high (solid diamond), very high (solid square). Key for species weights: weights are only shown for the taxa that most strongly followed the pattern in the PRC for clarity—the higher the weight (positive or negative), the more similar the response to that identified in the figure; taxa with negative weights show the opposite pattern to the PRC. An, animal, F, fungi, Ame, amoeba, Pl, plantae, Pro, protozoa, Rhi, rhizaria.

(*Micronuclearia*), a ciliate protozoa (*Zoothamnium*) and an ascomycete fungi from the class Sordariomycete. The MOTUs that most strongly followed a response opposite to that shown in the first PRC axis (i.e. were consistently negatively affected by the copper in the H and VH treatments) include three chlorophytes (two species of *Chlamydomonas* and one of *Selenastrum*), a chrysophyte (*Spumella* sp.), a bacillariophyte (*Navicula* sp.), two ciliate protozoa (of the subclasses Hypotrichia and Haptoria respectively), a Chytrid fungus (Chytridiomycota) and an arthropod (class Maxillopoda).

The second PRC axis highlights the response of MOTUs that, in the H and VH treatments, were initially different (greater number of reads) from those in the controls, but changed to be similar to the controls from 135 days onward (Fig. 3.4B). The MOTUs that most strongly followed the response shown in the second PRC axis and were positively affected by the copper in the H and VH treatments at the first two time points include several fungi (two Chytrid fungi [both of phyla Chytridiomycota], two ascomycetes [of the classes Leotiomycete and Sordariomycete, respectively], two basidiomycetes [*Resinicium* sp. and one of class Agaricomycetes] and two fungi of the order Entomophthorales [one belonging to the family Ancylistaceae and one to the family Ancylistaceae, of the genus *Conidiobolus*]) as well as two species of nematode within the family Monhysteridae, four amoeboid protozoa (*Flabellula* sp., *Saccamoeba*, one of the order Euamoebida and one of the order Centramoebida) and two Cercozoa (*Ebria* and one of class Thecofilosea). The MOTUs that most strongly followed a response opposite to that shown in the second PRC axis and were negatively affected by the copper present in the H and VH treatments at the first two time points include an ascomycete of the class Sordariomycete, an amoeboid protozoa (*Rhogostoma*) and one Cercozoa (*Cryothecomonas* sp.).

Although the eukaryote communities in the VL/L treatments appear to be similar to the C treatment in both the first and second PRC axes, slight differences in community composition between these three treatments can be seen in the nMDS plot (Fig. 3.1). There was a significant time × treatment interaction and significant time and treatment main effects ($p = 0.001$ for all). On all sampling occasions, there was a significant difference ($p < 0.05$) in the composition among treatments. At 15, 31, 135 and 161 days, all treatments were significantly different from the controls ($p \leq 0.04$). From 365 days onward, the VL treatment was similar to the C treatment ($p \geq 0.06$). From 407 days onward, the L treatment was also similar to the C treatment ($p \geq 0.06$).

Of the copper phases, sediment copper concentrations explained a greater proportion of the variation in the composition data (12.2%) than did the concentrations in the overlying water (11.2%) and pore water (8.5%). Overall however, time had the strongest influence (15.0%). Together, these variables accounted for 30.3% of the variation in the composition data, but of these variables, it was again time that had the greatest unique contribution to explaining assemblage composition (14.1%), compared to each of the sediment, pore water and overlying water concentrations which accounted for 2% or less each. The low unique contribution of the copper concentrations attests to their correlated nature ($r=0.68$–0.88) and together the copper contamination regime accounted for similar component of the variation in composition (15.3%) as did time.

4. DISCUSSION

Through DNA metabarcoding, the effect of copper on the composition of the whole eukaryote community in a freshwater environment has been assessed. By characterising a greater proportion of the biological community than a traditional analysis of the macroinvertebrate community over the same time period (similar to standard methods of bioassessment), subtle effects of copper at low levels of contamination have been observed. The results complement and increase understanding of the effects of copper gained from traditional types of assessment and provide opportunities to identify novel indicator species for toxicity studies with copper.

The compositional changes in the biological community observed were influenced by both time and copper concentrations. The influence of time reflects the initial colonisation and establishment of the community and then likely changes in the community due to seasonal variation in environmental factors (e.g. time and light). The greater influence of sediment copper concentrations over those of the overlying water and pore water in explaining the variation of the community composition reflects the benthic nature of the community, although the differences were small due to the co-correlation of the copper concentrations in the sediments, pore waters and overlying waters.

4.1. The biological response to copper

The biological response to copper concentrations characterised by DNA metabarcoding in this study was driven by meio- and microbiota, some of which were consistently, while others were only initially, positively or negatively affected by copper. Key drivers in the response included both autotrophs (chlorophytes, streptophytes, chrysophytes and bacillariophytes) and heterotrophs (fungi, nematodes and protozoa), demonstrating the importance of characterising the response of this fraction of the biological community in ecotoxicological assessments. However, there has been limited research into the effects of copper on the structure and function meio- and microbiota within freshwater communities (Massieux et al., 2004).

Studies that have considered the effects of copper on meio- and microbiota in the benthic layer have focused on phototrophic species within the biofilm. It is well known that photosynthesis and growth in algae is inhibited by copper at low concentrations (Garvey et al., 1991), and overall richness of phototrophic species has been shown to decrease in response to copper (Kaufman, 1982; Leland and Carter, 1984). In this study, three chlorophytes, one chrysophyte and one bacillariophyte were shown to be consistently sensitive to copper according to the PRC analysis (Fig. 3.4A). However in contrast, three chlorophytes and one streptophyte were tolerant to the contaminant, demonstrating variation of sensitivity amongst phototrophic species. Considering green algae further, within the chlorophytes, two MOTUs representing *Chlamydomonas* spp. were identified as sensitive to copper in this study, however there were 18 MOTUs identified as *Chlamydomonas* spp. in total; in other studies, *Chlamydomonas* spp. have been considered both tolerant (Knauer et al., 1997) and sensitive (Garvey et al., 1991) to the contaminant. One salient finding was that no bacillariophytes were identified to be positively affected by copper; this is consistent with previous research, which has shown green algae to dominate both plankton and periphyton communities in response to copper (Effler et al., 1980; Soldo and Behra, 2000).

Variation in sensitivity at the genus level can also be demonstrated by studying *Navicula*. Of the 11 MOTUs identified as *Navicula* spp., only one was identified as sensitive to copper in the PRC analysis (Fig. 3.4A). Species of *Navicula* have often been shown to be sensitive to metals, for example, in response to copper the growth of *Navicula incerta* was reduced (Rachlin et al., 1983). However, in this study, clearly some species were not sensitive to copper; this is consistent with community-level analyses of biofilms exposed to copper, where tolerant *Navicula* species have been identified (Guasch et al., 2002; Leland and Carter, 1985).

Although most studies that have considered the effects of copper on aquatic meio- and microbiota have focused on phototrophic species, one comprehensive study also considered taxonomic changes in heterotrophic species, specifically ciliate protozoa, within plankton exposed to copper (Le Jeune et al., 2007). They demonstrated a decrease in biomass of Haptorida, Prostomatida and Oligotrichida as a result of copper exposure (Le Jeune et al., 2007). In comparison, ciliate protozoa in this study were both consistently positively (*Zoothanmnium*) and negatively (two, of subclasses Hypotrichia and Haptoria, respectively) affected by copper (Fig. 3.4A), but there were many that were not affected, including one MOTU identified as Oligotrichia.

Only one fungus was identified as strongly and negatively responding to copper over the duration of the analysis; this was a Chytrid fungus (Fig. 3.4A). We assume, therefore, that copper did not particularly affect most fungi and, indeed, there were two fungal taxa, Protosporangium and Sordariomycetes, that were positively affected by copper. This tolerance of fungi to copper is consistent with a growing body of evidence that suggests some fungi can detoxify copper through the induction of genes encoding for metallothionein proteins (e.g. Ding et al., 2013). Copper can also increase the virulence of some fungal strains (Festa and Thiele, 2012) and may provide a competitive advantage for such taxa in the benthic environment.

The apparent copper tolerance of fungal taxa in our system contrasts that of fungal assemblages on leaf litter that were particularly affected by copper, evidenced by a reduction in the number of sporulating fungal species following copper exposure (Duarte et al., 2009). With this in mind, it was surprising that several fungi were identified that were driving the biological response to copper in the second PRC axis (i.e. during the initial establishment of the community), of which the majority were positively affected (Fig. 3.4B). This suggests that fungi were able to initially colonise the aquatic environment and thrive (comparatively) in the copper-contaminated mesocosms, whereas other organisms took longer to acclimate.

Two species of nematode, within the family Monhysteridae were also initial drivers of the community composition (PRC second axis), taking advantage while other organisms took longer to acclimate (Fig. 3.4B). Nematodes have previously shown sensitivity to copper (Boyd and Williams, 2003) and are often considered as potential indicators for biomonitoring due to their ubiquitous nature and interspecies variability in sensitivity (Bongers and Ferris, 1999; Boyd et al., 2000; Ekschmitt et al., 2001). However, there were 66 MOTUs classified as Monhysteridae, so considering that only two were key in driving the biological response, and that they only drove the response in the initial two sampling occasions, species within the family may not be suitable bioindicators for long-term copper contamination. However, if those MOTUs that initially responded positively could be identified at the species level (rather than family), they may be good indicators of a recent disturbance due to copper (or other) contamination.

5. IMPLICATIONS FOR BIOMONITORING

Metabarcoding enabled a far wider spread of biota to be examined than that which could be obtained using traditional optical techniques and a similar amount of effort. An analysis of the benthic macroinvertebrate community was performed over the same time period (Gardham et al., 2014a), which required the same sampling effort and approximately 50 days of sample processing and identification. Through this analysis, the response of 35 different macroinvertebrate taxa to copper was observed (Gardham et al., 2014a). In contrast, while a similar sample processing time was required for the DNA metabarcoding analysis (including method optimisation), 2202 MOTUs were distinguished, over 60 times the number of taxa identified by optical techniques. In fact, even when MOTUs were clustered at the phylum level, there was greater resolution compared to the optical analysis, with 75 different taxon clusters identified. The number of taxa identified by traditional optical analysis is within the range of standard processing of river biomonitoring samples, which typically results in the identification of between 10 and 100 families of macroinvertebrates (Baird and Hajibabaei, 2012). This benefit of DNA metabarcoding in characterising a greater proportion of the biological community compared to traditional techniques of biomonitoring has been recognised previously (Baird and Hajibabaei, 2012; Chariton et al., 2010; Hajibabaei et al., 2011).

With a greater proportion of the biological community taken into account compared to the macroinvertebrate analysis, the DNA metabarcoding analysis was more sensitive. Using DNA metabarcoding, significant differences in the biological composition of the benthic community among all individual treatments were observed, even the C (4.6 mg/kg dry wt; 1.5 μg/L pore water copper) and VL (71 mg/kg dry wt; 2.8 μg/L) treatments (apart from at the final two time points, where similar communities were found in the C/VL and C/VL/L treatments, respectively). In contrast, in the traditional optical analyses, only the C/VL/L treatments (<200 mg/kg dry wt; <5 μg/L) and H/VH treatments (>400 mg/kg dry wt; >18 μg/L) could be distinguished from each other (Fig. 3.5) (Gardham et al., 2014a). Chariton et al. (2010) were similarly able to discriminate differences in community composition between reference sites and sites with low levels of contamination using DNA metabarcoding.

The ability to identify sites with low levels of contamination using DNA metabarcoding is likely to greatly improve current biomonitoring practices. These practices are often unable to identify sites which are only moderately impacted (Chariton et al., 2010), because they are usually both time and resource limited, which prevents the incorporation of traditional techniques that would allow meio- and microbiota to be assessed. DNA metabarcoding has the potential to enable meio- and microbiota to be included in biomonitoring practices because it is less time intensive, does not require taxonomic expertise and will become cost-effective as sequencing technologies progress. The results in this study demonstrate that DNA metabarcoding analyses could provide sought-after evidence of moderately impacted sites, allowing management measures to be implemented earlier and preventing further contamination.

Further comparison of the invertebrate and DNA metabarcoding data sets indicates that the two methods of bioassessment should be used to support each other, rather than DNA metabarcoding replacing the traditional technique. First, although the invertebrate analysis identified nearly 40 × less taxa, many of the macrofauna identified by invertebrate analysis were not

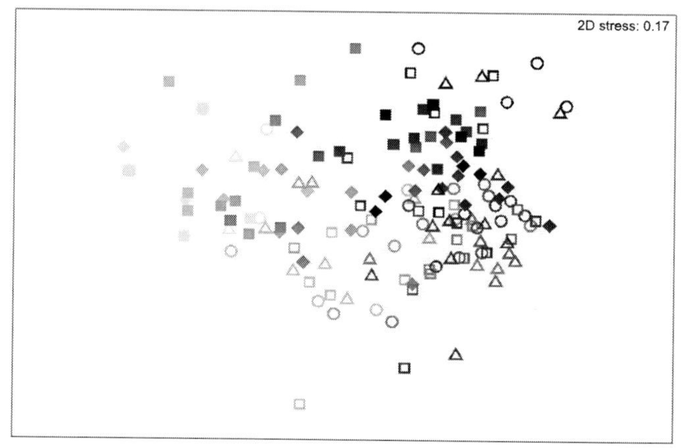

Figure 3.5 nMDS plot showing differences in community composition among invertebrate samples (Gardham et al., 2014a), where points closer together represent a more similar community composition than those further apart. Key: control (open square), very low (open triangle), low (open circle), high (solid diamond), very high (solid square) and gradient of light blue (light grey in the print version) to dark blue (dark grey in the print version) denotes change in time from 15 days through to 497 days.

found through DNA metabarcoding. This could be due to the relatively small amount of sample 10 g (around 10 mL) that the DNA was extracted from, compared to the samples processed for invertebrate analysis (100 mL of sediment each). Nevertheless, these larger invertebrates are valuable components of the ecosystem and their characterisation should not be lost from the bioassessment process.

For ecotoxicologists, DNA metabarcoding also presents opportunities to identify novel indicator taxa for traditional toxicity studies (Chariton et al., 2010). The MOTUs shown to respond to copper represented meio- and microbiota and some could be classified to the genus level (e.g. *Navicula, Rhogostoma, Flabellula*), providing accurate taxonomic information of possible indicator taxa that could be explored for toxicity studies. However, clearly there is variation in sensitivity even at the genus level and species level identification would be more informative. Nevertheless, a promising aspect of DNA metabarcoding for ecotoxicology studies is the identification of novel indicator taxa within the meio- and microbiota, that have previously been underrepresented, if considered at all, in bioassessment by toxicity assays.

5.1. Limitations

Despite its evident potential, DNA metabarcoding is still in the early stages of development and several limitations remain beyond the issue of obtaining abundance data. Firstly, bias in PCR and sequencing may obscure the presence of rare taxa, while conversely errors may generate artefactual sequences that could errantly be classified as rare taxa (Coissac et al., 2012). In addition, DNA can persist beyond the life span of an individual and even potentially disperse to new habitats in systems with high connectivity (e.g. rivers), confounding the bioassessment of solely live organisms at a particular site (Bohmann et al., 2014). However, some consider that DNA in dead organisms is normally rapidly degraded and only preserved under exceptional circumstances (Baldwin et al., 2013). Beyond sequencing and DNA persistence, a large investment is required to create high-quality taxonomic databases (Taberlet et al., 2012).

A key limitation that must be resolved is that there is no consensus for a single target marker region for eukaryotic molecular surveys (Meadow and Zabinski, 2012). Targeting of the gene encoding 18s rRNA has been shown to underestimate true species richness by Tang et al. (2012), who favoured the use of CO1. However, CO1 has poorly conserved priming sites (Tang et al., 2012). An individual marker with wide coverage across all eukaryote kingdoms is hard to find, so some studies have developed metabarcoding markers for several specific organism groups (e.g. Epp et al., 2012). This means multiple markers could be used to gain wider coverage of the biological community and more accurately capture the biodiversity. Previously, this use of multiple markers meant that sequencing power was vastly reduced; however, with the latest technology numerous genes can be analysed simultaneously at a sequencing depth suitable for ecological monitoring. Many of the issues in DNA metabarcoding will be resolved as technologies develop, for example, methods are currently being developed to remove the requirement of PCR, removing potential bias and error from that step (Taberlet et al., 2012).

5.2. Final conclusions

Ultimately, DNA metabarcoding is probably not the "silver bullet", but will add to the line of evidence rather than replace existing methods for biomonitoring (Yoccoz, 2012). It will enable the meio- and microbiota, at the base of the ecological food web, and critical to the biogeochemical functioning of the environment, to be routinely assessed. However, the charismatic megafauna are likely to be underrepresented by this technique, yet still need to be assessed and conserved (Baird and Hajibabaei, 2012). This study is one of the first to provide empirical evidence of the toxic effect of copper on the composition of the eukaryote community from the micro- to macroscale within a freshwater ecosystem, the first using DNA metabarcoding. It has highlighted the power of the technique and its potential for identifying novel indicator taxa for traditional toxicity studies. With more studies like this, we may not only be able to identify that an ecosystem has been altered and detrimentally affected by anthropogenic impacts, but also confidently identify the stressor that is causing the most severe impact.

6. DATA ACCESSIBILITY

The data from this project have been deposited in the Commonwealth Scientific and Industrial Research Organisation data access portal and are accessible at: http://dx.doi.org/10.4225/08/5428AFA7DF112.

ACKNOWLEDGEMENTS

Construction of the mesocosms was funded by a Macquarie University Infrastructure Grant. Stephanie Gardham was supported by a Macquarie University Research Excellence

Scholarship. CSIRO's Water for a Healthy Country flagship provided partial funding and specialist time. The authors thank A. Michie, M. Nagel and L. Oulton for their help in sample collection and two reviewers for their useful comments. The authors have no conflict of interest to declare.

APPENDIX

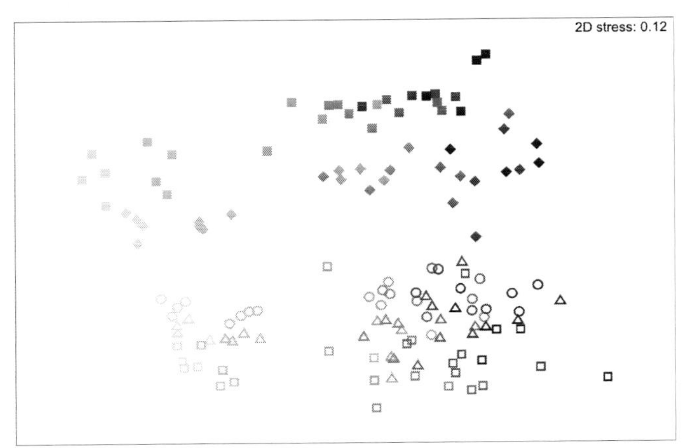

Figure 3.A1 nMDS plot showing differences in community composition among samples based on presence–absence data, where points closer together represent a more similar community composition than those further apart. Key: Control (open square), very low (open triangle), low (open circle), high (solid diamond), very high (solid square) and gradient of light blue (light grey in the print version) to dark blue (dark grey in the print version) denotes change in time from 15 days through to 497 days.

REFERENCES

Anderson, M.J., 2001. A new method for non-parametric multivariate analysis of variance. Austral Ecol. 26, 32–46.

Angly, F., Dennis, P., Skarshewski, A., Vanwonterghem, I., Hugenholtz, P., Tyson, G., 2014. CopyRighter: a rapid tool for improving the accuracy of microbial community profiles through lineage-specific gene copy number correction. Microbiome 2, 1–13.

Baird, D.J., Hajibabaei, M., 2012. Biomonitoring 2.0: a new paradigm in ecosystem assessment made possible by next-generation DNA sequencing. Mol. Ecol. 21, 2039–2044.

Baldwin, D.S., Colloff, M.J., Rees, G.N., Chariton, A.A., Watson, G.O., Court, L.N., Hartley, D.M., Morgan, M.J., King, A.J., Wilson, J.S., Hodda, M., Hardy, C.M., 2013. Impacts of inundation and drought on eukaryote biodiversity in semi-arid floodplain soils. Mol. Ecol. 1746–1758.

Blaalid, R., Carlsen, T., Kumar, S., Halvorsen, R., Ugland, K.I., Fontana, G., Kauserud, H., 2012. Changes in the root-associated fungal communities along a primary succession gradient analysed by 454 pyrosequencing. Mol. Ecol. 21, 1897–1908.

Bohmann, K., Evans, A., Gilbert, M.T.P., Carvalho, G.R., Creer, S., Knapp, M., Yu, D.W., de Bruyn, M., 2014. Environmental DNA for wildlife biology and biodiversity monitoring. Trends Ecol. Evol. 29, 358–367.
Bongers, T., Ferris, H., 1999. Nematode community structure as a bioindicator in environmental monitoring. Trends Ecol. Evol. 14, 224–228.
Boyd, W.A., Williams, P.L., 2003. Comparison of the sensitivity of three nematode species to copper and their utility in aquatic and soil toxicity tests. Environ. Toxicol. Chem. 22, 2768–2774.
Boyd, S.E., Rees, H.L., Richardson, C.A., 2000. Nematodes as Sensitive Indicators of Change at Dredged Material Disposal Sites. Estuar. Coast. Shelf Sci. 51, 805–819.
Bradford, T.M., Morgan, M.J., Lorenz, Z., Hartley, D.M., Hardy, C.M., Oliver, R.L., 2013. Microeukaryote community composition assessed by pyrosequencing is associated with light availability and phytoplankton primary production along a lowland river. Freshw. Biol. 58, 2401–2413.
Caporaso, J.G., Kuczynski, J., Stombaugh, J., Bittinger, K., Busman, F.D., Costello, E.K., Fierer, N., Pena, A.G., Goodrich, J.K., Gordon, J.I., Huttley, G.A., Kelley, S.T., Knights, D., Koenig, J.E., Ley, R.E., Lozupone, C.A., McDonald, D., Muegge, B.D., Pirrung, M., Reeder, J., Sevinsky, J.R., Turnbaugh, P.J., Walters, W.A., Widmann, J., Yatsunenko, T., Zaneveld, J., Knight, R., 2010. QIIME allows analysis of high-throughput community sequencing data. Nat. Methods 7, 335–336.
Chao, A., 1984. Nonparametric estimation of the number of classes in a population. Scand. J. Stat. 11, 265–270.
Chapman, P.M., Wang, F., Janssen, C., Persoone, G., Allen, H.E., 1998. Ecotoxicology of metals in aquatic sediments: binding and release, bioavailability, risk assessment, and remediation. Can. J. Fish. Aquat. Sci. 55, 2221–2243.
Chariton, A.A., Court, L.N., Hartley, D.M., Colloff, M.J., Hardy, C.M., 2010. Ecological assessment of estuarine sediments by pyrosequencing eukaryotic ribosomal DNA. Front. Ecol. Environ. 8, 233–238.
Chariton, A.A., Ho, K.T., Proestou, D., Bik, H., Simpson, S.L., Portis, L.M., Cantwell, M.G., Baguley, J.G., Burgess, R.M., Pelletier, M.M., Perron, M., Gunsch, C., Matthews, R.A., 2014. A molecular-based approach for examining response of eukaryotes in microcosms to contaminant-spiked estuarine sediments. Environ. Toxicol. Chem. 33, 359–369.
Chon, H.-S., Ohandja, D.-G., Voulvoulis, N., 2012. The role of sediments as a source of metals in river catchments. Chemosphere 88, 1250–1256.
Clements, W.H., 2000. Integrating effects of contaminants across levels of biological organization: an overview. J. Aquat. Ecosyst. Stress. Recover. 7, 113–116.
Clements, W.H., Cherry, D.S., Cairns, J.J., 1988. Structural alterations in aquatic insect communities exposed to copper in laboratory streams. Environ. Toxicol. Chem. 7, 715–722.
Coissac, E., Riaz, T., Puillandre, N., 2012. Bioinformatic challenges for DNA metabarcoding of plants and animals. Mol. Ecol. 21, 1834–1847.
Creer, S., Fonseca, V.G., Porazinska, D.L., Giblin-Davis, R.M., Sung, W., Power, D.M., Packer, M., Carvalho, G.R., Blaxter, M.L., Lambshead, P.J.D., Thomas, W.K., 2010. Ultrasequencing of the meiofaunal biosphere: practice, pitfalls and promises. Mol. Ecol. 19 (Suppl. 1), 4–20.
Deagle, B., Thomas, A., Shaffer, A., Trites, A., Jarman, S., 2013. Quantifying sequence proportions in a DNA-based diet study using Ion Torrent amplicon sequencing: which counts count? Mol. Ecol. Resour. 13, 620–633.
Ding, C., Festa, R.A., Chen, Y.-L., Espart, A., Palacios, Ò., Espín, J., Capdevila, M., Atrian, S., Heitman, J., Thiele, D.J., 2013. *Cryptococcus neoformans* copper detoxification machinery Is critical for fungal virulence. Cell Host Microbe 13, 265–276.

Duarte, S., Pascoal, C., Cássio, F., 2009. Functional stability of stream-dwelling microbial decomposers exposed to copper and zinc stress. Freshw. Biol. 54, 1683–1691.

Effler, S.W., Litten, S., Field, S.D., Tongngork, T., Hale, F., Meyer, M., Quirk, M., 1980. Whole lake responses to low level copper sulfate treatment. Water Res. 14, 1489–1499.

Egge, E., Bittner, L., Andersen, T., Audic, S., de Vargas, C., Edvardsen, B., 2013. 454 pyrosequencing to describe microbial eukaryotic community composition, diversity and relative abundance: a test for marine Haptophytes. PLoS One 8, e74371.

Ekschmitt, K., Bakonyi, G., Bongers, M., Bongers, T., Boström, S., Dogan, H., Harrison, A., Nagy, P., O'Donnell, A.G., Papatheodorou, E.M., Sohlenius, B., Stamou, G.P., Wolters, V., 2001. Nematode community structure as indicator. Eur. J. Soil Biol. 37, 263–268.

Epp, L.S., Boessenkool, S., Bellemain, E.P., Haile, J., Esposito, A., Riaz, T., Erséus, C., Gusarov, V.I., Edwards, M.E., Johnsen, A., Stenøien, H.K., Hassel, K., Kauserud, H., Yoccoz, N.G., Bråthen, K.A., Willerslev, E., Taberlet, P., Coissac, E., Brochmann, C., 2012. New environmental metabarcodes for analysing soil DNA: potential for studying past and present ecosystems. Mol. Ecol. 21, 1821–1833.

Festa, R.a., Thiele, D.J., 2012. Copper at the front line of the host-pathogen battle. PLoS Pathog. 8, e1002887.

Fleeger, J.W., Carman, K.R., Nisbet, R.M., 2003. Indirect effects of contaminants in aquatic ecosystems. Sci. Total Environ. 317, 207–233.

Gadd, G.M., Raven, J.A., 2010. Geomicrobiology of Eukaryotic Microorganisms. Geophys. J. Roy. Astron. Soc. 27, 491–519.

Gardham, S., Chariton, A.A., Hose, G.C., 2014a. Invertebrate community responses to a particulate- and dissolved-copper exposure in model freshwater ecosystems. Environ. Toxicol. Chem. 33, 2724–2732.

Gardham, S., Hose, G.C., Simpson, S.L., Jarolimek, C., Chariton, A.A., 2014b. Long-term copper partitioning of metal-spiked sediments used in outdoor mesocosms. Environ. Sci. Pollut. Res. 21, 7130–7139.

Garvey, J.E., Owen, H.A., Winner, R.W., 1991. Toxicity of copper to the green alga, Chlamydomonas reinhardtii (Chlorophyceae), as affected by humic substances of terrestrial and freshwater origin. Aquat. Toxicol. 19, 89–96.

Guasch, H., Paulsson, M., Sabater, S., 2002. Effect of copper on algal communities from oligotrophic calcareous streams. J. Phycol. 38, 241–248.

Hajibabaei, M., Shokralla, S., Zhou, X., Singer, G.a.C., Baird, D.J., 2011. Environmental barcoding: a next-generation sequencing approach for biomonitoring applications using river benthos. PLoS One 6, e17497.

Harrahy, E.A., Clements, W.H., 1997. Toxicity and bioaccumulation of a mixture of heavy metals in *Chironomus tentans* (Diptera: Chironomidae) in synthetic sediment. Environ. Toxicol. Chem. 16, 317–327.

Havens, K.E., 1994. Structural and functional responses of a freshwater plankton community to acute copper stress. Environ. Pollut. 86, 259–266.

Kaufman, L.H., 1982. Stream Aufwuchs accumulation: disturbance frequency and stress resistance and resilience. Oecologia 52, 57–63.

Kennedy, A.D., Jacoby, C.A., 1999. Biological indicators of marine environmental health: meiofauna—a neglected benthic component? Environ. Monit. Assess. 54, 47–68.

Knauer, K., Behra, R., Sigg, L., 1997. Effects of free Cu2+ and Zn2+ ions on growth and metal accumulation in freshwater algae. Environ. Toxicol. Chem. 16, 220–229.

Le Jeune, A.-H., Charpin, M., Deluchat, V., Briand, J.-F., Lenain, J.-F., Baudu, M., Amblard, C., 2006. Effect of copper sulphate treatment on natural phytoplanktonic communities. Aquat. Toxicol. 80, 267–280.

Le Jeune, A.-H., Charpin, M., Sargos, D., Lenain, J.-F., Deluchat, V., Ngayila, N., Baudu, M., Amblard, C., 2007. Planktonic microbial community responses to added copper. Aquat. Toxicol. 83, 223–237.

Leland, H.V., Carter, J.L., 1984. Effects of copper on species composition of periphyton in a Sierra Nevada, California, stream. Freshw. Biol. 14, 281–296.

Leland, H.V., Carter, J.L., 1985. Effects of copper on production of periphyton, nitrogen fixation and processing of leaf litter in a Sierra Nevada, California, stream. Freshw. Biol. 15, 155–173.

Maltby, L., 1999. Studying stress: the importance of organism-level responses. Ecol. Appl. 9, 431–440.

Massieux, B., Boivin, M.E.Y., Van Den, E.F.P., Langenskiold, J., Marvan, P., Barranguet, C., Admiraal, W., Laanbroek, H.J., Zwart, G., 2004. Analysis of Structural and Physiological Profiles To Assess the Effects of Cu on Biofilm Microbial Communities Analysis of Structural and Physiological Profiles To Assess the Effects of Cu on Biofilm Microbial Communities. Appl. Environ. Microbiol. 70, 4512–4521.

Mayer-Pinto, M., Underwood, A.J., Tolhurst, T., Coleman, R.A., 2010. Effects of metals on aquatic assemblages: what do we really know? J. Exp. Mar. Biol. Ecol. 391, 1–9.

McArdle, B.H., Anderson, M.J., 2001. Fitting multivariate models to community data: a comment on distance-based redundancy analysis. Ecology 82, 290–297.

Meadow, J.F., Zabinski, C.a., 2012. Spatial heterogeneity of eukaryotic microbial communities in an unstudied geothermal diatomaceous biological soil crust: Yellowstone National Park, WY, USA. FEMS Microbiol. Ecol. 82, 182–191.

Medinger, R., Nolte, V., Pandey, R.V., Jost, S., Ottenwälder, B., Schlötterer, C., Boenigk, J., 2010. Diversity in a hidden world: potential and limitation of next-generation sequencing for surveys of molecular diversity of eukaryotic microorganisms. Mol. Ecol. 19 (Suppl. 1), 32–40.

Morgan, M.J., Chariton, A.a., Hartley, D.M., Court, L.N., Hardy, C.M., 2013. Improved inference of taxonomic richness from environmental DNA. PLoS One 8, e71974.

Preston, B.L., 2002. Indirect effects in aquatic ecotoxicology: implications for ecological risk assessment. Environ. Manage. 29, 311–323.

Rachlin, J.W., Jensen, T.E., Warkentine, B., 1983. The growth response of the diatom Navicula incerta to selected concentrations of the metals: cadmium, copper, lead and zinc. Bull. Torrey Bot. Club 110, 217–223.

Roussel, H., Ten-Hage, L., Joachim, S., Le Cohu, R., Gauthier, L., Bonzom, J.-M., 2007. A long-term copper exposure on freshwater ecosystem using lotic mesocosms: primary producer community responses. Aquat. Toxicol. 81, 168–182.

Roussel, H., Chauvet, E., Bonzom, J.-M., 2008. Alteration of leaf decomposition in copper-contaminated freshwater mesocosms. Environ. Toxicol. Chem. 27, 637–644.

Serra, A., Guasch, H., 2009. Effects of chronic copper exposure on fluvial systems: linking structural and physiological changes of fluvial biofilms with the in-stream copper retention. Sci. Total Environ. 407, 5274–5282.

Serra, A., Corcoll, N., Guasch, H., 2009. Copper accumulation and toxicity in fluvial periphyton: the influence of exposure history. Chemosphere 74, 633–641.

Shaw, J.L., Manning, J.P., 1996. Evaluating macroinvertebrate population and community level effects in outdoor microcosms: use of in situ bioassays and multivariate analysis. Environ. Toxicol. Chem. 15, 608–617.

Shokralla, S., Spall, J.L., Gibson, J.F., Hajibabaei, M., 2012. Next-generation sequencing technologies for environmental DNA research. Mol. Ecol. 21, 1794–1805.

Soldo, D., Behra, R., 2000. Long-term effects of copper on the structure of freshwater periphyton communities and their tolerance to copper, zinc, nickel and silver. Aquat. Toxicol. 47, 181–189.

Swartzman, G.L., Taub, F.B., Meador, J., Huang, C., Kindig, A., 1990. Modeling the effect of algal biomass on multispecies aquatic microcosms response to copper toxicity. Aquat. Toxicol. 17, 93–117.

Taberlet, P., Coissac, E., Pompanon, F., Brochmann, C., Willerslev, E., 2012. Towards next-generation biodiversity assessment using DNA metabarcoding. Mol. Ecol. 21, 2045–2050.

Tang, C.Q., Leasi, F., Obertegger, U., Kieneke, A., Barraclough, T.G., Fontaneto, D., 2012. The widely used small subunit 18S rDNA molecule greatly underestimates true diversity in biodiversity surveys of the meiofauna. Proc. Natl. Acad. Sci. U. S. A. 109, 16208–16212.

Ter Braak, C.J.F., Smilauer, P., 2012. Canoco Reference Manual and User's Guide: Software for Ordination (Version 5.0). Microcomputer power, Itaca.www.canoco.com.

USEPA, 2013. ECOTOX User Guide: ECOTOXicology Database System (Version 4.0). Available: http:/www.epa.gov/ecotox.

Van den Brink, P.J., Ter Braak, C.J.F., 1999. Principal response curves: analysis of time-dependent multivariate responses of biological community to stress. Environ. Toxicol. Chem. 18, 138–148.

Willerslev, E., Davison, J., Moora, M., Zobel, M., Coissac, E., Edwards, M.E., Lorenzen, E.D., Vestergard, M., Gussarova, G., Haile, J., Craine, J., Gielly, L., Boessenkool, S., Epp, L.S., Pearman, P.B., Cheddadi, R., Murray, D., Brathen, K.A., Yoccoz, N., Binney, H., Cruaud, C., Wincker, P., Goslar, T., Alsos, I.G., Bellemain, E., Brysting, A.K., Elven, R., Sonstebo, J.H., Murton, J., Sher, A., Rasmussen, M., Ronn, R., Mourier, T., Cooper, A., Austin, J., Moller, P., Froese, D., Zazula, G., Pompanon, F., Rioux, D., Niderkorn, V., Tikhonov, A., Savvinov, G., Roberts, R.G., MacPhee, R.D.E., Gilbert, M.T.P., Kjaer, K.H., Orlando, L., Brochmann, C., Taberlet, P., 2014. Fifty thousand years of Arctic vegetation and megafaunal diet. Nature 506, 47–51.

Yoccoz, N.G., 2012. The future of environmental DNA in ecology. Mol. Ecol. 21, 2031–2038.

INDEX

Note: Page numbers followed by "*f*" indicate figures, "*b*" indicate boxes and "*t*" indicate tables.

I

L

M

N

O

P

Q

R

S

V

W

ADVANCES IN ECOLOGICAL RESEARCH VOLUME 1–51

CUMULATIVE LIST OF TITLES